P9-AQV-206

PERGAMON INTERNATIONAL LIBRARY
of Science, Technology, Engineering and Social Studies
*The 1000-volume original paperback library in aid of education,
industrial training and the enjoyment of leisure*
Publisher: Robert Maxwell, M.C.

TIME AND MAN

*"What then is time? If no-one asks me, I know; if I
am asked to explain it, I do not know."*

St. Augustine.

THE PERGAMON TEXTBOOK
INSPECTION COPY SERVICE

An inspection copy of any book published in the Pergamon International Library will
gladly be sent to academic staff without obligation for their consideration for course
adoption or recommendation. Copies may be retained for a period of 60 days from
receipt and returned if not suitable. When a particular title is adopted or recommended for
adoption for class use and the recommendation results in a sale of 12 or more copies, the
inspection copy may be retained with our compliments. The Publishers will be pleased to
receive suggestions for revised editions and new titles to be published in this important
International Library.

Other titles of interest

ALSBERG: In Quest of Man: A Biological Approach to the Problem of Man's Place in Nature

BROWN: By Bread Alone
(An analysis of the changing world food situation)

DUNCAN & WESTON-SMITH: Encyclopaedia of Ignorance (2 vols)
(An account of what lies beyond the edge of our knowledge in the Life, Earth and Physical Sciences).

GEORGE: Machine Takeover: The Growing Threat to Human Freedom in a Computer - Controlled Society

SCHAFF: History and Truth
(Describing the differing and sometimes contradictory representations of historical events passed down by historians)

STEG: Should We Limit Science and Technology?

Nude descending staircase, No. 2

by

Marcel Duchamp

An artist's attempt to catch the two aspects of time,
permanence and change.

TIME AND MAN

by

L.R.B. ELTON

and

H. MESSEL

PERGAMON PRESS

OXFORD · NEW YORK · TORONTO · SYDNEY · PARIS · FRANKFURT

U.K.	Pergamon Press Ltd., Headington Hill Hall, Oxford OX3 0BW, England
U.S A.	Pergamon Press Inc., Maxwell House, Fairview Park, Elmsford, New York 10523, U.S.A.
CANADA	Pergamon of Canada Ltd., 75 The East Mall, Toronto, Ontario, Canada
AUSTRALIA	Pergamon Press (Aust.) Pty. Ltd., 19a Boundary Street, Rushcutters Bay, N.S.W. 2011, Australia
FRANCE	Pergamon Press SARL, 24 rue des Ecoles, 75240 Paris, Cedex 05, France
FEDERAL REPUBLIC OF GERMANY	Pergamon Press GmbH, 6242 Kronberg-Taunus, Pferdstrasse 1, Federal Republic of Germany

First edition 1978

L.C.# 76-26511

Library of Congress Cataloging in Publication Data
Elton, Lewis Richard Benjamin.
Time and man.
Bibliography: p.
Includes index.
1. Time. I. Messel, Harry, joint author.
I. Title.
BD638.E44 1976 115 76-26511
ISBN 0-08-021332-4 (Hardcover)
ISBN 0-08-021331-6 (Flexicover)

In order to make this volume available as economically and rapidly as possible the authors' typescript has been reproduced in its original form. This method unfortunately has its typographical limitations but it is hoped that they in no way distract the reader.

Printed in Great Britain by William Clowes Co. Ltd., Beecles, Suffolk.

CONTENTS

Contents

Contents

PREFACE

We have called this book *Time and Man*, for although it is only
in the final chapter that we are specifically concerned with the
impact of time on ourselves, all chapters reflect the endeavours
of men to probe the mysteries of time and to elucidate its properties. The book is aimed at students of all ages and we hope
to have met the needs and interests of all, whether their bias
is on the arts or science side. Being ourselves scientists, we
have taken science as our starting point, but the subject
stretches across the physical sciences, the biological sciences
and the humanities, and we have endeavoured to maintain a proper
balance between these. The text is both philosophical and
factual in nature, and we set ourselves the task to give an
account of our subject, without the introduction of any mathematics, beyond the very simplest algebra.

Factual information is presented not for its own sake, but to
help to lead to a better understanding of the concepts associated with time. We wish to stress strongly here that science
does not consist solely of a host of unrelated properties of
matter or of nature; nor is the main objective of science just
to discover as many new phenomena as possible. Surely the purpose must be to understand these phenomena and properties in
terms of the minimum possible number of basic laws.

Thus we endeavour to present the subject material in a fashion
which will make the student THINK and QUESTION and at the same
time realize that science is still a wide-open subject, with
many of the most difficult questions usually not even being
broached, let alone being answered. This is important in this
present day when one so often hears and reads that the most
interesting part of science has been completed, that only the
i's remain to be dotted and t's crossed. Nothing could be
further from the truth, as the reader may appreciate after
reading this book.

The subjects covered in the book range so widely that no single
person could be expert in them all. We are very conscious of
this, and we shall be very grateful for any corrections to be
pointed out to us.

Preface

We have provided a list of references for further reading and referred to some of the books and articles in several of the chapters of this book.

Our thanks are due to the many people, from pupils at school to Professors who have read parts of the manuscript and greatly improved it. These include Professor L. Castillejo, Professor V. L. Ehrenberg, Miss Bridget Elton, Mr. L. H. Hawkins, Miss Christine Holdsworth, Mr. D. L. Hurd, Mr. D. E. James, Mrs. Morag Morris and Dr. A. W. Wilson. The idea of the book was conceived during a visit of one of us (H.M.) to England on a Commonwealth Visiting Professorship, and much of the book was written while the other (L.R.B.E.) visited Sydney with the support of the Science Foundation for Physics. We gratefully acknowledge this help.

We are indebted to Faber and Faber Ltd., London, and Harcourt Brace Jovanovich Inc., New York, for permission to reprint extracts from T. S. Eliot's Collected Poems 1909 - 1962 and Murder in the Cathedral, and to the Philadelphia Museum of Art for permission to reproduce Marcel Duchamp's Nude descending a staircase, No. 2.

L. R. B. Elton and H. Messel

Guildford and Sydney, 1976.

Chapter 1

THE EXPERIENCE OF TIME

Our own experience

The passage of time is something familiar to all of us. The days come and go, and with each passing day we grow older. In this highly mechanized world of today, innumerable clocks and watches ceaselessly tick away time and determine the schedules by which hundreds of millions of people live. The passage of time is thus a concept which is second nature to us, and which all of us understand thoroughly, or do we?

Few of us are likely to remember a time when we were not aware of the passage of time. We recall the occasions when it hung heavily on our hands, when we were bored and short of things to do, and the occasions when it passed all too quickly, because we were interested. And yet it cannot always have been so. It seems likely that a sense of time is not innate, but develops early as a result of everyday experience common to us all; in other words, it results from learning. It steals up on us unawares, and becomes part of us before we are conscious of it. In this, it is no different from many other matters that to us seem obvious beyond question, until we start to question them. It is one of the purposes of this book to question what we mean by time; to question whether the questions we may formulate are meaningful; and if they are, to look for answers.

Before we embark on this programme, it will be useful to rehearse the experiences which we associate with the concept of time. As infants we may have been fed on demand or to a strict time-table, and we experienced the succession of day and night. In this way we were exposed to more or less regular rhythmic changes, which were reflected in rhythmic changes in our own bodily condition, to be hungry or sated, awake or asleep. *Time is associated with rhythm and change.* Soon after, we began to move and the speed with which we crawled determined the time we took to get from one end of the playpen to the other. *Time is associated with speed and velocity.*

1

We grew older, learnt to speak and listened to stories. "Once
upon a time ..." was how they started, and we became aware of
the fact that things happened before we were born. Time is
associated with history, with going backwards as well as forwards, with the succession of events. Soon we were made to be
punctual and perhaps were punished or reprimanded for being
slow or late. We may have taken kindly to this or not, but in
any event it contributed to the formation of our character.
Time is associated with social behaviour.

By now we had become conscious of time and began to ask questions. Where was I before I was born? What happens after
death? What did God do before he created the world? *Time is
associated with the great questions of philosophy and religion.*
We learnt to read time on the clock and in this modern world,
where the clock is ubiquitous and our lives are governed by it,
we came perhaps to identify time with the clock. *Time is associated with something outside us over which we have no control,
something that appears absolute.*

At this point we may for the first time have formulated the
question: "What is time?".

What is time?

We have purposely described the growing-up process and its
interaction with the concept of time, as if it were the common
experience of us all. This is clearly too simple, and we hope
that you, the reader, may have reflected at times and questioned
whether this was indeed so in your case. If you have begun to
question, then we have succeeded in putting you into the frame
of mind appropriate for the subject under discussion - one that
doubts and questions, and does not accept uncritically. We
have further omitted all references to the findings of those
who have studied the mind and particularly the minds of children
and who may have reached quite different conclusions from the
ones presented above. We shall learn about these later but at
present we assume that you have not studied them and are merely
trying to answer that question "What is time?" in the light of
your own experience, as you perceive it, or perhaps as we have
tried to perceive it for you.

What conclusions can we draw from our short discussion? Firstly,
that we shall always be influenced by earlier ideas, so that we
have prejudices, of which we are mostly unaware. The question-
ing frame of mind is designed to help identify such prejudices

so that we can be conscious of them. Secondly, that we associate time with many and very diverse experiences, to an extent that it becomes doubtful whether there is one concept that is "time". If this is so, then the very idea that there is one thing that is time is a prejudice and false, and the question "What is time?" turns out to be based on this false prejudice.

At this point, time appears to have eluded us, and we obviously must stalk it more carefully. Perhaps we should start by investigating how to ask the right kind of question. We shall attempt to do this in the next chapter.

Chapter 2

THE QUESTIONS OF SCIENCE

Collection of data - what?

Scientific activity usually starts with the collection of observable phenomena within a given field. Some would call this activity prescientific - Lord Rutherford, the famous physicist, called it stamp collecting - but even stamp collectors are not satisfied with the mere amassing of quantities of stamps. They classify them into groups according to some general principle; country of origin, colour, picture, etc., or a combination of these. Which classification is preferred depends on its usefulness under particular circumstances. Thus scientists might classify water as a liquid, as a conductor of electricity or as a necessity of life. Now the process of classification depends on the principle that all entities are in one sense unique and in another sense similar to other entities. By assigning a given entity to a sufficient number of different classes it is generally possible to characterize it uniquely, and to that extent we can answer the question "What is gold?" by saying that gold is a yellow metal that does not tarnish in air. Unfortunately, this at once leads to the next question, e.g. "What is a metal?", and the answer to that question defines a class. It is easy to see that this series of questions can be continued until - since there is only a finite number of words in the language - we must come to a point where we try to define something in terms of something else that we met earlier on and are still trying to define, at which point we appear to have reached a logical circle. Gertrude Stein expressed this dilemma most succinctly in her famous definition of a rose: "A rose is a rose is a rose." However, having said this, it must be conceded that "What" questions, the answers to which are either classifications or definitions, are useful in science, as experience shows. However, when it comes to time, the problem appears to be that whatever time is, it is essentially unique, and that any similarity to other entities is somewhat superficial. We may say that time flows and include it among the entities that flow, but this does not justify us in applying the laws of hydrodynamics to it. Thus we again conclude tentatively that not

only is there no single concept which is time, which makes the
"is" in "What is time?" inappropriate, but that whatever time
is is so nearly unique as to make the "What" inappropriate too.

Description and correlation of data - how and why?

The next usual step in scientific enquiry is the interpretation
of observed phenomena. This leads to "How" questions - "How do
the planets move?" "How do plants grow?" - and the answers lead
to the association of the new observation with earlier ones,
with which we are more familiar. To give an example from a
mythological age, the ancient Greeks interpreted the observation
of the motion of the Sun across the sky in terms of a flaming
chariot, driven by the sungod Helios. The point here is that
chariots and flames were familiar everyday objects to them in a
way that the distant Sun was not and never could be. Today we
interpret the same phenomenon in terms of the gravitational
force between the Sun and the Earth, where force is a concept
with which we are familiar through our muscles. In general,
"How" questions eventually lead to concepts that are so familiar
that we stop asking about them, but this is just where danger
lies, for these very familiar concepts are exactly those which
we met at a very early age and which form our commonsense view
of the world. That this can be a false view has already been
indicated in the previous chapter.

A good deal of this book will be concerned with a description
of phenomena associated with time and how these relate to other
concepts and phenomena. Beyond this, there are "Why" questions,
which relate phenomena to the great unifying laws of nature and
eventually lead to the building of scientific theories. Of
these, we shall also have something to say.

Model making

In general, "How" questions look for the similarities between
the entity being investigated and more familiar entities. This
search for entities which in certain respects are like the
entity being investigated has proved very fruitful and is called
the building of a model. By its very nature, a model in some
ways is like the entity of which it is a model, and in other
ways is unlike it. Thus we think of water as a fluid when we
are concerned with the motion of ships, but as a collection of
electric charges when we are concerned with electroplating.
Both the fluid and the collection of electric charges are models
of water and each shares different aspects of its properties.

Their use enables us to systematize in each case a range of
phenomena associated with water and to predict phenomena previously unknown, which have to be checked against experiment,
but it is obvious from the fact that we use more than one model,
that each model also has its limits, beyond which it is not
valid. Taking these considerations over to the concept of time,
we conclude that "What is time like?" is a perfectly legitimate
question, which could lead to the building of models, as long
as we remember that each model can only represent certain aspects
of time.

All this was expressed with clarity by W. C. D. Whetham as early
as 1903:*

> "The function of science is merely to construct
> a consistent set of phenomena. Whether that
> model corresponds with the ultimate reality
> behind phenomena, whether indeed there be any
> ultimate reality, is a question for metaphysics,
> not for Natural Science. An imaginary model
> of the whole of Nature would be too complex
> for the mind to grasp."

Pointer readings

By now it should be clear that one of our problems will be to
make sure that our questions are meaningful and that mere gram-
matical correctness is not an adequate criterion for this. In
science we ask questions of nature in the form of experiments,
and the answer often comes in pointer readings on dials, i.e.
the answers are numerical. It therefore becomes necessary *for
this purpose* to define time precisely in numerical terms, and
the obvious way to do this is through the pointer reading of a
clock. When one first reads that statement, one may well ask:
"Is time no more than what a clock measures?" "Is time not
something absolute that exists whether we *have a clock to
measure it or not*?" It may well be that it was one of Einstein's
great discoveries that from the point of view of scientific
enquiry, the only way we can handle time - or indeed any other
entity - is through the numbers associated with its measurement,
which in turn is defined through the measuring instrument. Such
a definition of time is called an *operational definition*, and

---

*E. Homberger (Ed.), *The Cambridge Mind*, Jonathan Cape, 1971,
   p. 175.

Einstein showed in particular that a definition of time as something absolute and independent of a measuring instrument - so to speak, in terms of ideal minutes, but not of measured minutes - led to contradictions with experiment. Of course, we need not think of the clock in this connection as necessarily a device with a dial and hands. We shall see that there are many processes in nature which measure the passage of time, but we must be careful always to relate this passage of time to one of these natural processes or else to a man-made measuring device. In either case, we can eventually arrive at a pointer reading.

It must be realized, however, that the operational definition of time in terms of a measuring device will enable us to investigate only certain aspects of time, and it is these which we say are amenable to scientific investigation. Conversely, we shall find that the operational definition leads to the conclusion that certain questions are in principle unanswerable, i.e. they will not yield a measurement that gives an answer to the question. Such questions are then meaningless not only in the context of scientific investigation, but usually in that of investigations of any kind. Finally, in line with Whetham's statement above, there are many areas of human thought and endeavour, where the discipline of science appears inappropriate, and we shall trace certain aspects of the concept of time through language, literature, philosophy and religion. These aspects are no less real to men than those investigated by science, in fact to many the quantification of time that is inherent in the scientific approach appears cold and inhuman. We shall endeavour to present a view which pays due regard to other ways in which men have looked at time, and in this way hope to show that as scientists we know that the world is more than science and that we are glad of it.

Chapter 3

THE MEASUREMENT OF TIME

Time interval and point of time

We now turn to a discussion of the devices that have been used
to measure time. Even in the restricted operational definition
of time, it must be realized that the word has two distinct,
although related, meanings. One is that of duration, *an interval
of time;* the other is that of a specified instant of time, *a
point in time.* They are related, because a point in time may
be identified as being the end of the time interval which started
at some arbitrary but fixed reference point in time, such as the
founding of Rome or the birth of Christ. Thus the question
"What is the time?", which clearly refers to a point in time,
is answered by, say, "10 a.m.", which is the time interval since
a certain reference point in time, that in this case is midnight,
last night. Nevertheless, conceptually the two meanings are
very different and we must be careful not to confuse them.*
Measuring devices determine intervals of time, although our
clocks and watches are designed to read directly points in time.
For this to be possible, they have to be standardized against a
standard clock, which is known to measure accurately the time
interval up to the present point in time from the fixed reference point. Eventually, this must take us back to one universal
standard clock, against which all other standard clocks are calibrated. This universal standard clock then defines operational
time in both its meanings of time interval and point in time.

Basic properties of clocks

The above discussion makes it clear how time can be defined
operationally through pointer readings on a measuring instrument, but it tells us nothing about the properties that this
instrument must possess. It cannot be stressed too strongly

---

*Look back to the first sentence of the second paragraph of
 Chapter 1, where the first use of "time" refers to a point in
 time, and the second to a time interval.

that we are not talking here about some ideal measuring device,
but that the operational definition requires us eventually to
construct a real instrument, made of real materials, with all
the limitations inherent in real instruments.

Although our measuring device, when constructed, will define
time, we need our *intuitive ideas about time* in order to specify
the properties of the instrument. There is no way to get round
this circular conclusion, which in no sense is unique to the
problem of defining time. To give an example, we define weight
operationally as a pointer reading on a weighing machine and
we can do this because there is general agreement regarding what
is meant by weight and hence the properties that a good weighing
machine must have. We also define a person's intelligence
quotient as the score which he obtains in an appropriate intelligence test, but as there is no general agreement as to what is
meant by intelligence, there is also no general agreement as to
what constitutes a good intelligence test.

The two most common images of time are those in which time is
likened to an infinite line and to an ever-flowing uniform
stream. The former is a spatial analogy, and we have used it
already in the expression "point in time", since point is a
spatial concept. However, the spatial analogy is a dangerous
one, since in principle it is possible for a person to visit all
points in space, while in time it is possible only to visit
points in the future and not in the past. As a model, in the
sense that we discussed this concept in the previous chapter,
space is therefore not very useful. The image of the ever-
flowing stream is very much better, since it combines the properties of an infinite line and of a sense of direction along
it. Taking this model a stage further, we can think of time as
a substance contained in a reservoir, from which the stream is
fed. The uniform decrease of the quantity of time in the reservoir is then a measure of the flow of time. This model we can
actually build and an example of it is the familiar hourglass
in which sand trickles steadily from an upper to a lower reservoir. In this way we have built an instrument that measures
time and which at the same time (note how one cannot get away
from the use of the word "time" in transferred meanings!) is a
model of what we imagine time to be like. This in itself ensures
the necessary consensus of opinion that we have found an appropriate device for measuring time.

Although we may imagine time as a uniform ever-flowing stream,
we do not in fact really experience it in this way. The aspect

of it which is quite clear to us is the direction in which the
stream flows. Intuitively we never appear in any doubt as to
the direction in which time flows, but the rate at which it
flows is perceived by us far less precisely. Experiments on
people who have spent long periods in very uniform surroundings,
say in deep caves, have shown that such people often grossly
misjudge the duration of time which they have spent in these
surroundings - they lose their "sense of time". To maintain
this sense, it is necessary to reinforce it through the experience of rhythmic change, such as the alternation of day and
night. Our experience of time is therefore closely associated
with that of rhythmic change and we measure for instance the
duration of a long time interval in terms of the number of day
and night alternations contained in it. A measuring device
which incorporates a rhythmic change is therefore a model not
of time itself or of how we imagine it, but of how we indirectly
experience its rate of flow. A pendulum clock is obviously an
example of such a measuring device.

Standards of time

In practice most time-measuring devices, which we shall call
clocks, have been of the one or other of the above two kinds
and the problem has been to ensure in the first kind, in which
there is a permanent change, that this change is uniform, and
in the second kind, in which the change is rhythmic, that the
time of duration or time interval of each rhythm remains constant. In principle, this requires a further and more accurate
device, clearly an impossibility when the first device is itself
to be the standard. Throughout all our history man has had two
obvious and natural standards of time, the day and the year.
The day is an easy period of time to recognize because we
experience daylight and darkness as a result of the spin of the
Earth on its axis. The year also has always been a relatively
easy period of time to recognize because of the seasons - caused
by the tilting of the Earth's axis. Speaking in a general way,
we say a day is the period of time for one complete revolution
of the Earth about its axis, and the year is the time taken for
the Earth to make one complete revolution around the Sun. It
does this in just under 365¼ days. We shall sharpen up these
definitions of time later in this chapter.

It is completely natural, therefore, that firstly the day and
secondly the year should be used by man to measure the passage
of time. For accurate time measurements during a day, the duration of each day has been divided by 24 to give us hours; these

in turn are divided by 60 to give us minutes and these again by
60 to give seconds.

Standard reference points in time

Having discovered how to standardize time intervals, we next
have to find a method for standardizing certain reference points
in time. At a very early age, men fixed the middle of each day
as the moment when the Sun was at its highest point during that
day. This we call noon and adjust our clocks to read 12 when
it occurs. However, if it is noon at one place, it will be
midnight at the opposite place on the Earth with different times
at other places. It is clear, therefore, that the actual time
as shown by clocks will be different at different points on the
Earth's surface.

In order to see in detail how times vary, it is convenient to
draw circles around the Earth passing through the Poles, as
indicated in Fig. 3.1.

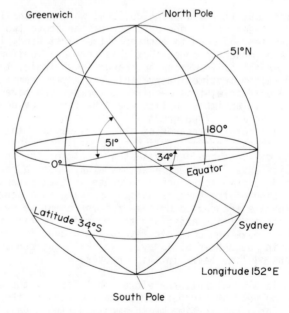

Fig. 3.1. Lines of longitude and latitude.

Each circle is called a meridian. Let us compare two meridians such as, for example, the one passing through Greenwich, the observatory near London in England, and the one passing through Sydney - as shown in Fig. 3.1. If one swung the meridian through Greenwich around to Sydney we would be moving it eastward through 152°. This is described by the term *longitude* and Sydney is said to have a longitude of 152° east of Greenwich. It is clear that, with the Earth rotating so that Sydney is moving eastward towards the right-hand side of Fig. 3.1, Sydney will see the Sun at an earlier time than Greenwich by an amount depending on the 152° of longitude. In fact we can easily work out that, since one complete revolution of 360° corresponds to one day, which is 24 hours or 24 x 60 minutes, every degree of revolution of the Earth corresponds to

$$\frac{24 \times 60}{360} = 4 \text{ minutes of time.}$$

Thus, Sydney must be 152 x 4 = 608 minutes or just over 10 hours ahead of Greenwich in time. By tradition Greenwich has been taken to have zero degrees longitude and the longitude of all other places on the Earth is given as so many degrees east or west of Greenwich. It was F. G. W. Struve (1793 - 1864), the director of the famous Pulkovo observatory near Leningrad, who took the decisive step towards the international recognition of Greenwich as the prime meridian. Places are said to be east of Greenwich if they have a longitude less than 180° east of Greenwich and the times of these places are ahead of that at Greenwich. In this way one-half of the Earth's surface has longitude east of Greenwich. The other half of the Earth's surface has longitude of anything up to 180° west of Greenwich, and the times of all these places lag behind that of Greenwich.

If one starts from the longitude at Greenwich, therefore, and works eastward one can readily work out how far ahead of Greenwich any place is in time. A place which has a longitude of 45° east is 3 hours ahead of Greenwich, a place which is 90° east has time 6 hours ahead of Greenwich; and finally a place which is almost 180° east is 12 hours ahead of Greenwich in time.

Similarly, if we work back the other way, 45° west of Greenwich means 3 hours behind Greenwich time; a longitude of 90° west signifies a time 6 hours behind Greenwich; and finally, a location which is almost 180° west of Greenwich has a time almost 12 hours behind that of Greenwich.

There is only one problem with all this, which can be seen from
the following example. Suppose, say, it is 2 a.m. on a Sunday
morning at Greenwich and we work out the times of other places
on the Earth. As we go eastward we finally reach a point on
the meridian 180° from Greenwich which will be 12 hours ahead,
and this will have a time of 2 p.m. on the Sunday. On the other
hand as we go westward from Greenwich the time will drop further
and further back until the point when we are almost on the meridian 180° west and the time will be 2 p.m. on the Saturday.
Thus, we can reach a certain spot on the 180° meridian which
would be 2 p.m. Sunday if we travelled to it eastward from
Greenwich and 2 p.m. Saturday if we travelled to it westward
from Greenwich. The time of the day would be the same but the
day would be different. Thus the 180° meridian directly opposite the Earth from the Greenwich meridian is called the International Date Line. If a plane or ship travelling eastward
crosses this line the day suddenly changes back one - for example,
from Sunday to Saturday, or from Friday to Thursday. On the
other hand, if the ship or aircraft is travelling westward and
crosses the Date Line the day will suddenly make a jump of one
day forward.

This gives rise to such interesting effects as a traveller being
able to have two birthdays. If he crosses the International
Date Line travelling eastward say at 2 a.m. on Sunday morning
it will suddenly become 2 a m. Saturday morning and he will
have Saturday all over again. On the other hand, a traveller
crossing in the other direction will find 2 a.m. on a Saturday
morning suddenly becomes 2 a.m. Sunday morning and he will appear to have missed the Saturday almost entirely.

The fact that local time differs from place to place became an
inconvenience as soon as travel became reasonably fast, but it
is surprising to realize that as recently as 1840, different
towns in England kept different times, so that early railway
time-tables looked like modern air time-tables. A Great Western
time-table of 1841 states that*

> "London time is kept at all Stations on the
> Railway, which is about 4 minutes earlier than
> Reading time, 5½ minutes before Steventon time,
> 7½ minutes before Cirencester time; 8 minutes
> before Chippenham time; and 14 minutes before
> Bridgwater time."

---

*Quoted in Lawrence Wright, *Clockwork Man*, Elek, London, 1968,
p. 145.

Rail travel in due course led to the need for a rationalization
of the chaos of local times and an international agreement was
reached in 1884, to divide the Earth into time zones approximately 15° of longitude wide and based on lines of longitude
15° apart, starting with 0° at Greenwich. With minor variations,
due to local conditions, this system is now universal. It is
shown in Fig. 3.2.

It is, of course, true that knowing the longitude east or west
of Greenwich of a given place does not tell one exactly where
the place is on the Earth's surface; it can be anywhere on a
particular semicircle extending from pole to pole. Another
angle is required in order to completely specify a given point.
This is called the angle of latitude. The latitude of Sydney,
for example, is 34° south as shown in Fig. 3.1. In effect this
is measured in the following way:

Imagine firstly a line from the centre of the Earth to that
point on the Equator with the same longitude as Sydney; secondly
imagine a line drawn directly from the centre of the Earth to
Sydney. The angle between these lines is close to 34° south of
the Equator.

In the early days of navigation sailors had considerable trouble
in determining their positions. To locate themselves on the
globe they had to determine both latitude and longitude. The
former could be obtained by measuring the height of the Sun at
midday, provided one knew what day of the year it was. It was
done by means of an instrument called a sextant and was known
as "shooting the Sun". However, the determination of longitude
proved difficult for many years. For an accurate determination
of longitude one needs accurate clocks. Consider, for example,
a ship setting out from England with its clocks set at Greenwich
time. After, say, several months of voyaging the longitude can
be immediately determined by noting from the Sun when it is 12
noon and seeing how many hours and minutes the clocks differ
from 12 o'clock. Since each degree of longitude causes a time
difference of 4 minutes, the precise longitude east or west of
Greenwich is thus known.

The difficulty, however, was that in the early days clocks could
not be constructed accurately enough for good longitude deter-
minations. This was so much of a problem that in 1714 a reward
of £20,000 was offered by the British Government for any means
of determining a ship's longitude accurately to within about
half a degree. The reward was won by John Harrison, a self-

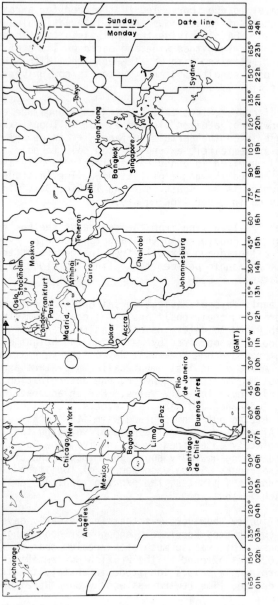

Fig. 3.2.   The division of the earth into time zones.

taught Yorkshire carpenter, who invented a marine time-keeper
which fulfilled the conditions for the reward. You may realize
what a step forward this was in the 18th century from the fact
that a clock fulfilling the British Government's condition had
to keep time accurately to better than 3 seconds per day - a
standard which, at the time when the reward was offered, had
not even been achieved by the best pendulum clocks on shore.
The name *chronometer* was given to these accurate clocks for
navigation, and today, of course, extremely accurate chronometers
are used.

<u>Making clocks</u>

Let us return to the construction of time-measuring devices.
It is not the purpose of this book to give detailed historical
accounts of the development of clocks, but a brief account of
the principles is indicated. The earliest clocks were undoubtedly sundials of various kinds and these were followed by devices
which used the regular flow of a substance such as water, oil
or sand, or the steady combustion of oil or candles. The earliest of these date from about 1600 B.C. in Egypt, and they were
used throughout classical times and the Middle Ages. In their
most sophisticated form, the outflow of water drove a mechanical
movement. This in turn led to the idea of weight or gravity-
driven clocks, in which a weight, at the end of a string that
unwound from an axle, drove the clock mechanism. Such clocks
could not be successful as long as the weight descended continuously, for it then did so with increasing speed. What was needed
was a check which would allow the weight to descend only a very
small distance before it was stopped and then to start again.
Here for the first time we see the introduction of the idea of
a rhythmic motion and the problem was first solved through the
invention, in about the 13th century, but by whom is not known,
of the escapement. The ingenuity of the original solution and
its development by some of the greatest scientists in subsequent
centuries are such that they are still worth our attention today.

The escapement is shown diagrammatically in Fig. 3.3. In it a
sawtoothed crown wheel, so called because the teeth are at
right angles to the plane of the wheel as in a crown, is driven
by the weight, but held in check by one of the two so-called
pallets which are fixed to an axle, known as the balance staff,
which rotates at right angles to the crown wheel. As the crown
wheel starts to move, it releases the upper pallet and turns the
balance staff with its associated balance weight until the lower
pallet engages the crown wheel and momentarily stops it. As

the lower pallet engages at the diametrically opposite end of
the crown wheel to where the upper pallet had engaged, the relative motion of a wheel and pallet is now in the opposite direction. Hence the balance staff starts rotating in the opposite
direction and the crown wheel moves on until stopped again by
the upper pallet. In this way a stop-start motion is achieved.

The main weakness of the escapement in its simple form is that
its motion is due entirely to the impact of the teeth of the
crown wheel on the pallets, and that this does not lead to a
very uniform oscillation of the balance staff. An important
development in clock construction was the introduction of the
pendulum; its principle was first discovered by Galileo in 1581.

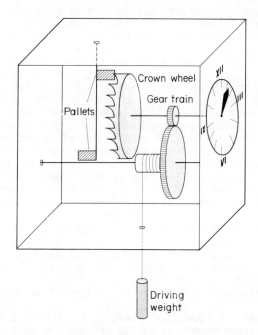

Fig. 3.3. The earliest form of escapement for
a clock.

The pendulum of a clock merely consists of an object or bob
attached to the lower end of a light rod, the upper end of which
is suspended from a support in such a manner that the rod and
its bob are free to swing to and fro under the influence of

gravity. The fact noted by Galileo is that the time for one
complete to and fro swing of such a pendulum is almost completely
independent of the magnitude of swing as long as this is not too
large. Thus a swinging pendulum will have almost exactly the
same time of swing whether oscillating through say 15° or 5°.
The duration of swing of a pendulum varies with its length and
a pendulum which makes one swing per second is almost exactly
1 metre long.

The following story of how Galileo came to study the pendulum
is told. In 1581, at the age of 17 years, while kneeling in the
Cathedral in Pisa, he observed the swinging of the Great Cathedral lamp and, using his own pulse beat to measure the time,
noted that the time of swing of the lamp was constant. He wrote
afterwards:

> "Thousands of times I have observed vibrations,
> especially in churches, where lamps, suspended
> by long cords, had been inadvertently set into
> motion...But I never dreamed of learning...
> (that each) would employ the same time in
> passing..."

Thus Galileo discovered that a swinging pendulum would regularly
"tick" away a unit of time - the period of swing - and that,
even if the swing gradually died down, the period would remain
unaffected. From this observation many types of pendulum clocks
were developed based on the swinging pendulum, and in all of
them only a very small amount of energy was needed to keep the
pendulum swinging. A quite small weight attached to a string
and being pulled down gradually by gravity was enough. Such
pendulum clocks were much more accurate than previous types and
many forms of pendulum clocks are still in existence today.
Huygens (1629-1695) was the first to use a pendulum to control
the motion of the escapement discussed above and an improved
version, the anchor escapement, was invented by William Clement
some fifty years later and is shown in Fig. 3.4.

The main subsequent improvements were the replacement of the
weight by a driving spring and of the pendulum by a balance
spring, and of compensatory devices that allowed for the consequences of changes of temperature. The first clock, driven by
electricity, was invented as early as 1843, but the next really
significant development was the quartz-crystal clock. In this
use is made of the property of a quartz-crystal that, when set
in mechanical vibration, it produces an alternating electric

potential difference between opposite faces. (This is called
the piezoelectric effect.) The vibration, which is of very
high frequency, may be used to control the frequency of an
alternating electric current which drives the clock. Such
clocks can be accurate to one part in ten million. Finally, we
come to the atomic clock, in which the frequency of a certain
spectral line emitted by caesium atoms is utilized to provide a
reference frequency. The accuracy of even early versions of
this clock as judged by a comparison of several different such
clocks of the same type, was of the order of one part in a hundred thousand million, i.e. it would not gain or lose more than
one second in 3000 years. Such accuracy is needed in many of
the experiments involving satellites and space rockets. Recent
clocks have been a hundred times more accurate still.

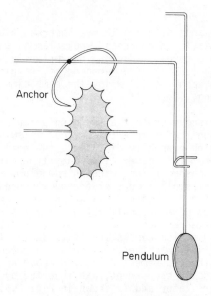

Fig. 3.4. The anchor type of escapement which
made it possible to incorporate a
pendulum in a clock.

Solar time

We have said previously that it is 12 noon for any observer when
the Sun passes most nearly overhead - that is, when the Sun is

vertically above his meridian. The time between the successive
crossings of the Sun over the same meridian is called a day - a
solar day.

So far this sounds very simple. It appears an excellent way
of measuring time. Scientifically, however, this definition of
time has its problems, because the day defined in this way turns
out to be not exactly constant. The solar day varies slightly
in length throughout the year, and we endeavour to make our
clocks tick through 24 hours in what is the average time of a
solar day throughout the year.

The following experiment demonstrates that the solar day is not
quite the same at different times of the year. Figure 3.5 is a
diagram to illustrate what happens, for example, if there is a
hole in a roof which permits a spot of sunlight to fall on the
floor of a room. If this spot is marked each day at noon, when
the sun is highest, then after a year a figure eight will have
been traced on the floor with its long direction running north
and south. The north-south movement of the spot of light is,
of course, nothing to worry about; it is due to the fact that
the height of the sun at noon differs at different times of the
year.

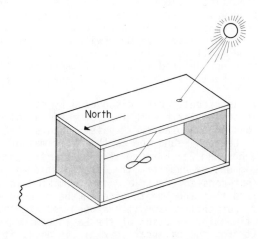

North

Fig. 3.5.   Experiment to show that the solar
day is not of constant length.

The east-west movement of the spot of sunlight, however, indicates that the Sun is sometimes slightly ahead of our clocks and
sometimes slightly behind them.

One reason for this stems from the fact that the Earth's orbit
around the Sun is not exactly circular - it is slightly elliptical. The Earth's speed around this orbit changes slightly -
when the Earth is closest to the Sun the speed is greatest and
when the Earth is furthest away the speed is least. Thus the
distance travelled by the Earth in its orbit in 24 hours changes
slightly over the course of a year. The time taken for the
Earth's spin to point a certain meridian at the Sun a second
time is sometimes slightly in excess of 24 hours and sometimes
slightly less. This effect of the Earth's elliptical orbit,
coupled with the tilt of the Earth's axis of spin, produces the
figure eight of Fig. 3.5. The Sun can sometimes be up to a
quarter of an hour ahead of our clocks, and sometimes as much
as a quarter of an hour behind them. This, of course, has little
practical importance in our everyday living.

For scientific purposes, however, it can be of importance that
we have a definition of a day which is not variable. For that
reason, the real Earth and its motion was replaced for time-
keeping purposes by an imaginary Earth which revolved round the
Sun in a circle with uniform speed once a year, and in this way
averaged out the fluctuations in speed of the real Earth. Because this speed was now an average or mean, the corresponding
time was called "mean time", and because the time measurement
was based on the observatory at Greenwich, the time measured in
this way was and still is called *Greenwich Mean Time*. The solar
day is therefore now defined in terms of the spin of this
imaginary Earth about its axis and not of the real Earth.

Even here, it has recently been discovered by comparing the
period of spin of the Earth with that of a number of other
periodic motions in the planetary system, such as the time of
revolution of the Moon round the Earth, the times of revolution
of Mercury, Venus and Earth round the Sun and the times of revolution of the moons of Jupiter, that the rotational spin period of
the Earth is not constant, possibly due to motion in the liquid core
of the Earth, so that the length of even the mean solar day is
not constant. For that reason it has been replaced by the year,
which is the time of revolution of the Earth round the Sun. To
distinguish it from the calendar year of 365 days, it is called
the tropical year.* Its length is 365.242199 solar days, and

*The word *tropical* here refers to its original derivation from
 the Greek word trope (solstice) and has nothing to do with a
 hot climate.

the difference between these two numbers is the reason for the
necessity of leap years. Time based on the length of the tropical year is called ephemeris time, from the Greek word for
calendar, and the second as the unit of time is defined through

$$1 \text{ tropical year} = 31,556,925.9747 \text{ seconds.}$$

Most recently atomic clocks have been used to define our standard of time and these are now so accurate that in due course
differences between atomic time and ephemeris time may become
apparent. Further, the atomic time of an event can be obtained
at the time of the event, so that in practice atomic time has
already replaced ephemeris time.

The calendar

A problem that has exercised man for a long time is that of a
simple and efficient calendar. The difficulty lies in the fact
that the three obvious periodicities, due to the rotation of the
Earth, the revolution of the Moon about the Earth and the revolution of the Earth about the Sun, are not simply related. In
fact,

$$1 \text{ tropical year} = 365.2422 \text{ solar days,}$$

while the length of the month, i.e. the observed time between
one full moon and the next, is given by

$$1 \text{ month} = 29.5306 \text{ solar days.}$$

At various times, people have counted by months, by calendar
years, or by mixtures of months and years, which last method
can be very confusing. Even if we ignore the Moon, we still
have the problem of the solar year not being a whole number of
days, which is dealt with by a system of leap years.

Our calendar owes something to many people and still bears traces
of their influence. These include the Babylonians, Egyptians,
Romans, the Christian Church and probably many others. No wonder we are left with a complicated system, that children and
even adults find difficult to remember. Our purpose, however,
is not to discuss the best method of making a calendar, but to
illustrate, through our present calendar, some of the foibles
of mankind. Thus we owe the idea of a Leap Year to Julius
Caesar, who also decreed that months should alternately have 30
and 31 days, except for February, which at that time was the

last month of the Roman year. (That is why our ninth month is
called September, etc.) All would have been well, if it had
not been decided to call the fifth month Julius in his honour.
For when Augustus followed him, he too wanted a month and chose
the next one. Too late did he realize that his month was shorter
than Caesar's and he promptly took a day from February, added
it to his month and then adjusted the rest of the year, so that
there would not be three long months in succession. And so because of the vanity of Emperor Augustus, to this day children
chant:

"Thirty days has September ....".

Attempts over the last fifty years to rationalize the calendar
by international agreement through the League of Nations and
the United Nations have uniformly failed. One is reminded of
the great Emperor Charles V, who ruled half the world but, over-
taken by melancholy, resigned from all his mighty positions, to
retire to a monastery, where he occupied himself with trying
to make a number of clocks show the same time. He failed, and
in despair asked himself how someone who could not make a few
clocks agree could hope to rule men.

Chapter 4

THE OPERATIONAL DEFINITION
OF TIME

Introduction

Towards the end of Chapter 2 we stated that if time was to be
investigated through the methods of science then it would have
to be defined through pointer readings on appropriate measuring
devices. In the following chapter we discussed the problem of
designing and building such devices and we were thrown back on
to our intuitive knowledge of time in order to specify properties that such a device had to possess. However, once we have
agreed on these, the device defines time, even if it turns out
that in certain circumstances time then has properties that
were not known to us before or are even in contradiction with
our intuitive knowledge. It is the purpose of this chapter to
illustrate this perhaps rather cryptic statement by investigating certain phenomena normally associated with time, using the
operational definition. Any conclusions that we reach must of
course be verified through experiment before they can be accepted.

Measuring velocity

In our discussion of the measurement of time in the previous
chapter, we saw how the concept of motion appeared inextricably
involved with that of time; in fact the two appear so intimately
related that we talk naturally of them without being aware of
each separately. In turn, our awareness of motion is usually
manifested through speed or velocity and hence it is natural
for us to investigate this phenomenon using the operational
definition of time. Throughout this book we shall, as is often
done, be using the terms speed and velocity interchangeably,
even though in scientific language the term velocity is reserved
for speed in a given direction.

We thus begin by investigating the measurement of velocity.
The average velocity of an object, travelling in a straight
line from A to B, is obtained by measuring the distance travelled,
and the time taken over the distance, and dividing the one by
the other. We shall assume for the present that there is no

85431

particular difficulty in measuring the distance accurately,
with the help of a calibrated metre stick. Let us then see how
we can obtain an accurate measure of the time interval. The
simplest method might be to attach a clock to the object and
read off the time elapsed as it goes from A to B. To make sure
that the measurement was accurate, we could compare our clock
with a standard clock before and after it has been used in the
experiment. This would have to be done with the clock at rest,
while in the experiment the clock was in motion. Could we be
sure that the clock would keep the same time in motion and at
rest? The answer is that we could only do this by sending signals from our moving clock to the stationary standard clock,
since the two would not be in the same place. Now no signal is
instantaneous and so an allowance has to be made for the time
taken for the signal to travel from the one clock to the other.
This in turn requires us to know the signal velocity and so we
must conclude that in order to measure one velocity, we must
already have been able to measure another!

The above argument leads us to reject the use of a moving clock.
If we use stationary clocks, then clearly we require two, one
at A and one at B, and these have to be carefully synchronized.
This can be achieved in two ways. In the first, the clocks are
synchronized at A and one of them is then taken to B. The drawback of this method is that, if the rate at which the two clocks
tick is not the same, or if the rate is affected by taking the
clock from A to B, then by the time the clocks are in position
at A and B, they will no longer show the same time. Further,
however small the difference in the two rates, the resulting
discrepancy in synchronization increases continually with time.
This was the problem faced by navigators in the days before
radio, who had to rely on the accuracy of the ship's chronometer
over long periods of time in their determination of longitude.
Nowadays chronometers are checked regularly against time signals,
broadcast by the radio stations, and this is in fact the second
method of synchronization, in which the clocks are positioned
at A and B, and synchronization is then achieved by sending a
signal from A to B, giving the time of the clock at A at fre-
quent intervals. The clock at B can then be adjusted accor-
dingly. Unfortunately, this method, which is indeed the only
practical one if the distance AB is large and a high degree of
accuracy is required, suffers from the drawback, which we have
met before, that no signal is instantaneous.

We conclude that it is not possible to measure a velocity with
complete accuracy, and that this is not due to deficiencies in

the measuring apparatus, but to intrinsic contradictions in the
situation. In practice, we minimize the error by using signals
with velocities very much larger than the velocity which we are
measuring, and in general we use light or radio waves. Their
velocity is so unimaginably large - it has been measured to be
about 300,000 kilometres per second - that until very recently
it could for all practical purposes be taken as infinite, so
that signals were assumed to be transmitted instantaneously.
This is certainly acceptable for ships at sea, but equally certainly is not acceptable for signals sent, say, to space probes
that are exploring Mars and Venus. Incidentally, the finiteness
of the velocity of light has also become a limitation in computer
construction. Computers are now so fast that they can switch
an electrical impulse in one thousand millionth of a second,
during which time light travels only 30 cm. To avoid delays,
computers have to be built compactly.

Eventually, therefore, we are still left with the problem of
investigating the velocity of light. In order to avoid the
difficulty of the synchronization of clocks in different places
while we are measuring this velocity, this is usually done by
reflecting a light pulse in a mirror, and so making it travel
to and fro over the same distance. While the time taken can
now be measured on one stationary clock, the experiment only
gives us the average velocity of light, say from A to B and back
from B to A, and there is no way of finding out whether in fact
the velocity was the same in both directions. After all, we
know in the case of sound that the answer would depend on whether
the air was still or not, but there is nothing corresponding to
the air for light, which travels in a vacuum. We therefore
still have no exact way of determining a signal velocity between
A and B and thus cannot synchronize our two clocks.

One possible way out of this difficulty might be to have a
third clock at C, the mid-point of AB and to send out signals
from C in both directions to A and B. Then, although we may
not know the signal velocity, it is possible to synchronize
clocks at A and B, provided that we know that the signal velocity is the same in both directions. But as we just pointed
out, this is exactly what we do not know.

Simultaneity and absolute time

Let us take this matter one stage further. If we wish to know
whether two events in different places took place at the same
time, then we are faced with exactly the same difficulty, that

we could only do this if we could have synchronized clocks in
the two places, and we know that this is something we cannot
have. And since what we in principle cannot determine, cannot
be said operationally to have a meaningful existence, the question "Did the two events at A and B take place simultaneously?"
is not a meaningful question.

Why then have we come to this conclusion, when in fact we would
all say that, from our own everyday experience, we conclude that
two events *can* happen simultaneously? The explanation is that
from our everyday experience we would also conclude incorrectly
that the velocity of light is infinite and that light signals
therefore are transmitted instantaneously. In that case, of
course, simultaneity does exist.

The commonsense view that we can speak of two events, which may
be separated by considerable spatial distances, as happening
simultaneously, has also quite logically led to the idea of an
absolute time, i.e. a time which does not depend on any natural
events or, in the words of Isaac Newton, (1642-1727) "of itself,
and from its own nature, flows equally without relation to anything external". This idea of time, which is deeply ingrained
in us, by its very definition is unobservable, and yet is logically required if we are to make a statement regarding the
simultaneity of events separated by a distance. As we have just
shown, no such statement can be made operationally and we therefore have no need for the hypothesis of an absolute and universal
time.

The realization that the traditional concept of absolute time
was irrelevant to observational science was due to Leibnitz, a
contemporary of Newton. The fact that it was only possible to
determine time locally, relative to particular observers was,
however, one of Einstein's great insights, and it led him eventually to postulate his theory of relativity, to be discussed
in the next chapter, which has revolutionized our conception of
the natural world. The local time referred to is the time as
measured on the clock held by a particular observer, and is
usually referred to as his proper time.

We have shown in this chapter that, at a deep conceptual level,
our ideas of time, as derived from commonsense experience, are
open to serious doubt, and we shall in the next chapter see
where Einstein's insight leads us. However, the practical
results of these changes in our ideas occur very largely at
velocities close to that of light, which can only be reached in

sensitive laboratory experiments. Even astronauts reach only
about one ten-thousandth of the velocity of light, and so, apart
from the next chapter, we shall be concerned with effects observed at the kind of velocities with which we are familiar.

Chapter 5

TIME AND RELATIVITY

Velocity of light

We now return to the problem raised in the previous chapter, as
to how we must modify our ideas of time to take into account
the fact that the velocity of light, though very large, is
finite.

The velocity of light through vacuum - usually designated by
the symbol c - is one of the most fundamental quantities in
science. To be more precise, it turns out that c is the velocity
of propagation not only of light but also of the electric field;
the fact that this is finite then gives rise to what are called
electromagnetic phenomena. It also follows that with c being
the basic velocity of propagation of the electric field, precisely this same velocity must automatically be the velocity
with which all electromagnetic radiation travels. It does not
matter whether we are talking about electromagnetic radiation
of radio frequencies, light frequencies or X-ray frequencies,
it must always travel with this basic velocity c.

As is to be expected, the first measurement of this velocity
was actually obtained by observing light, and hence the name
"velocity of light" has been retained for c despite the fact
that it is also the velocity of all other types of electromagnetic radiation.

It is interesting that the first even approximate measurement
of the velocity of light was done not on Earth, but astronomi-
cally and involved the planet Jupiter and its moons. Jupiter
is orbiting the Sun a distance about five times further out
than the Earth. It is a very large planet with a mass more than
three hundred times that of the Earth. One of the most interesting things about Jupiter is its system of twelve known moons.
The four brightest were seen by Galileo (1564 - 1642), using
the recently invented telescope, in 1610 and since then eight
other smaller moons of Jupiter have been observed. The four
bright satellites of Jupiter are comparable in size to our own

31

Moon, and the two largest are actually bigger than the planet
Mercury. But for their close proximity to Jupiter we would be
able to see at least three of them easily with the naked eye,
and they can readily be discerned with field glasses.

The periods of revolution of these four moons range from one
day 18 hours for the one closest to Jupiter, up to 16 days 16
hours for the one furthest out. Their orbits are quite close
to the plane of the ecliptic, which is the plane containing the
orbits of the Earth and Jupiter around the Sun, and therefore
each of these moons disappears from our view behind Jupiter for
a time once in each complete revolution. This is called an
eclipse of the moon in question. It was in 1675 that the Danish
astronomer Ole Roemer reported observations on the eclipses of
Jupiter's moons which allowed him to estimate the velocity of
light through vacuum.

He consistently observed the eclipses of one of Jupiter's four
bright moons over the period of a year starting from a time
when the Earth and Jupiter were as close to each other as possible. He obtained the, at first, surprising result that the time
taken for a certain number of eclipses to occur during the first
six months was slightly longer than for the same number of
eclipses during the second six months. This he interpreted as
being due to a finite velocity of propagation of the light
coming from the particular moon under observation.* The Moon
itself, of course, must always keep moving with a constant
period of revolution; thus the effect must only be an apparent
one and not due to any real fluctuations in period of the Moon
itself. Roemer's argument can be followed through in the following manner:

If the Earth and Jupiter are at their closest approach to each
other at a certain time, then close to six months later they
will be separated by the greatest possible distance, since the
period of Jupiter (11.9 years) is much greater than that of the
Earth. The difference between the two distances is essentially
equal to the diameter of the Earth's orbit around the Sun (see
Fig. 5.1). Then, while the Earth is moving away from Jupiter
(during the first six months), eclipses apparently occur later
and later than they would if the Earth and Jupiter had remained
in the same position; this is due to the extra time taken by
the light to cross the increasing gap between the two planets.

---

*We observe all planets and moons in the solar system by means
of light from the Sun being reflected from them to us.

Roemer found the time delay between observations at positions E_1 and E_2 to be 16.5 minutes. Since the diameter of the Earth's orbit was not very accurately known at the time, the value for the velocity of light which he obtained - about 200,000 miles per second - was about 10 per-cent too large, according to modern measurements. It did however show conclusively that the velocity of light was not infinite.

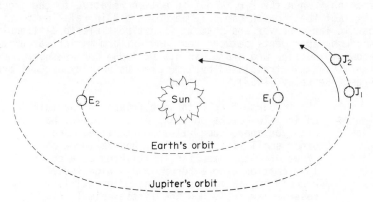

Fig. 5.1. This diagram (not to scale) illu-
          strates how the velocity of light
          was measured by Roemer.

In the years following this first observation more and more accurate methods of measuring the velocity of light have been devised. Foucault devised a reasonably accurate laboratory method in 1850. He obtained the value c = 298,000 km/sec. Between 1926 and 1933 a very accurate series of laboratory measurements was performed by Michelson and collaborators, who obtained the result c = 299,796 ± 4 km/sec for the velocity of light in vacuum.

By now a whole series of extremely accurate measurements have been performed not only on light but also at other wavelengths of electromagnetic radiation, and the accepted value at present is c = 299,792.458 ± 0.001 km/sec. This makes the velocity of electromagnetic radiation the most accurately determined number in physics. Every accurate observation has confirmed that this velocity of propagation of electrical influences through vacuum is a fundamental constant of nature, and all electromagnetic radiation is propagated through a vacuum with this speed.

The theory of relativity

We have so far overlooked a rather important point and that is
that velocities must be measured relative to something. Thus
for an aircraft in flight we measure both its ground speed and
its air speed, i.e. the speed of the aircraft relative to the
ground and relative to the surrounding air, and the two are the
same only when the air is not in motion relative to the ground.
The fact that motion can only be described in relation to a
particular observer and that it is not possible to distinguish
through experiments between rest and uniform motion was already
known to Galileo who wrote, in his *Dialogue Concerning Two Chief
World Systems* (translated by S. Drake, University of California
Press, Berkeley, 1953):

> "Shut yourself up with some friend in the main
> cabin below decks on some large ship, and have
> with you there some flies, butterflies, and
> other small flying animals. Have a large bowl
> of water with some fish in it; hang up a bottle
> that empties drop by drop into a wide vessel
> beneath it. With the ship standing still,
> observe carefully how the little animals fly
> with equal speed to all sides of the cabin.
> The fish swim indifferently in all directions;
> the drops fall into the vessel beneath; and,
> in throwing something to your friend, you need
> throw it no more strongly in one direction than
> another, the distances being equal; jumping with
> your feet together, you pass equal spaces in
> every direction. When you have observed all
> these things carefully (though there is no
> doubt that when the ship is standing still
> everything must happen in this way), have the
> ship proceed with any speed you like, so long
> as the motion is uniform and not fluctuating
> this way and that. You will discover not the
> least change in all the effects named, nor
> could you tell from any of them whether the
> ship was moving or standing still. In jumping
> you will pass on the floor the same spaces as
> before, nor will you make larger jumps toward
> the stern than toward the prow even though the
> ship is moving quite rapidly, despite the fact
> that during the time that you are in the air

the floor under you will be going in a direction
opposite to your jump. In throwing something
to your companion, you will need no more force
to get it to him whether he is in the direction
of the bow or the stern, with yourself situated
opposite. The droplets will fall as before
into the vessel beneath without dropping toward
the stern, although while the drops are in the
air the ship runs many spans. The fish in
their water will swim toward the front of their
bowl with no more effort than toward the back,
and will go with equal ease to bait placed anywhere around the edges of the bowl. Finally,
the butterflies and flies will continue their
flights indifferently toward every side, nor
will it ever happen that they are concentrated
toward the stern, as if tired out from keeping
up with the course of the ship, from which they
will have been separated during long intervals
by keeping themselves in the air."

When Galileo wrote this passage and for 200 years afterwards
these considerations were applied only to the mechanical motion
of material objects. Then, in the middle of the nineteenth
century, Maxwell (1831-1879) developed his theory of electro-
magnetism, which linked electrical and optical phenomena and
first showed that the velocity of light is a fundamental constant relating also to electromagnetic phenomena. This in turn
yielded the realization that if electromagnetic phenomena were
to be described in the same relative manner as was generally
accepted for mechanical phenomena, then *this led eventually and
indeed inevitably to the conclusion that their velocity, i.e.
the velocity of light, was independent of the motion of the
observer who measured it,* and thus was a universally constant
quantity.

This conclusion may have been inevitable, but it took the genius
of Einstein (1879-1955) to achieve it, in a famous paper published in 1905. To see why it is so difficult to accept it,
let us draw the analogy to the case of sound. As we know, sound
propagates with a certain velocity through the atmosphere of the
Earth, which we shall call V. This velocity is a property of
the atmosphere itself and is related to the normal kinetic
motion of the atoms and molecules within the atmosphere. Thus
the velocity of sound in air is V *relative to the air.*

If a train which is stationary blows its whistle, the sound
propagates through the air at the velocity V, relative to the
air. A stationary listener some distance d away hears the sound
at time $t = d/V$ after the whistle was blown (assuming a still
atmosphere).*

If instead the train had been moving at the time it blew its
whistle a stationary listener at some distance d away would still
hear the sound at a time $t = d/V$ later, because the sound propagates through the atmosphere with velocity V, quite independent
of how the source was moving at the time. As is well known, the
apparent frequency heard by the listener changes (a phenomenon
known as the Doppler effect), but the velocity of propagation
of the sound through the atmosphere is a constant. The Doppler
effect is illustrated in Fig. 5.2.

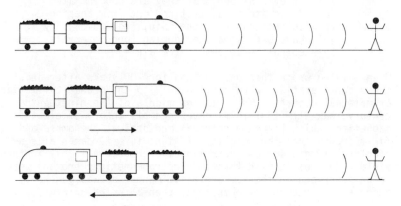

Fig. 5.2.   The pitch of the train whistle
            depends on whether the train is
            stationary or moving towards or
            away from the listener. This
            illustrates the Doppler effect.

---

*We have used here the elementary fact that since distance =
velocity multiplied by time,

$$\text{time} = \frac{\text{distance}}{\text{velocity}}$$

However, if the listener himself is moving, the time interval
will be different. For example, if a listener moves towards a
source of sound with velocity v, the sound will, of course,
then approach him with the velocity $V + v$. The listener is, in
fact, "running into the sound" and will therefore hear it sooner
than he would have had he remained stationary. Similarly, if
the listener moves away from the source of sound with speed v,
the sound only catches up on him at the *relative speed $V - v$*.

Thus with sound the speed that is a constant quantity is the
speed of sound V with respect to the atmosphere. Any observer
at rest with respect to the atmosphere will naturally observe
that the speed of sound is V. But if an observer is moving at
speed v with respect to the atmosphere he will observe a speed
of sound equal to $V + v$, for sounds that are emitted in front
of him, or $V - v$ for sounds that are emitted behind him; in
general, therefore, the speed of sound which he actually observes
relative to himself depends on his own motion through the atmosphere.

One difference between light and sound is that for the latter
we can refer the motion to the propagating medium, air, which
is quite independent of the observer. There is no corresponding
medium for light, and the velocity of light can therefore only
be referred to the observer. (Scientists in the nineteenth
century thought that there *must* be a medium and called it the
ether. In spite of many efforts, no evidence of the ether has
ever been found.) This leads to Einstein's conclusion that the
velocity of light is the same for any and every observer.
Furthermore, if this conclusion is right, then an observer who
moves towards a source of light, one who is at rest relative to
the source and one that moves away from it, all obtain the same
value for the velocity of light. This is in complete contradiction to the corresponding experiment with sound, and against all
common sense. There we had three values for the velocity,
$V + v$, V, $V - v$. For light, we only have one, c.

Whether Einstein is right in postulating that the velocity of
light is a universal constant can only be verified by experiment. Of the many experiments that have been performed in connection with the postulate, the most famous one, by Michelson
and Morley, was actually performed twenty years before Einstein's
paper. Its purpose at the time was to detect the motion of the
Earth through the hypothetical ether by measuring the velocity
of light in two opposite directions. If the Earth was moving
through the ether, then on the analogy with sound, the velocity

should be different in the two directions. No such difference
was found, which at the time was a very puzzling result. A
recent and very spectacular verification of the constancy of
the velocity of light depends on the properties of the so-called
neutral pions, which are elementary particles that arise as a
result of nuclear reactions. These pions can exist for only a
very short time and then decay spontaneously into gamma-rays,
which are a form of electromagnetic radiation of very short
wavelength. In 1964, Alvä, Farley, Kjellman and Wallin measured
the speed of such pions produced in the large particle accelera-
tor at Geneva and found it to be 99.75 per-cent of the speed of
light. In spite of this enormous speed, the speed of the resulting gamma-rays relative to the laboratory was equal to c within
an accuracy of 0.01 per-cent. If the normal rules for velocities
had held, the speed as measured in the laboratory should have
been the sum of the speed of the pions and that of the gamma-
rays relative to the pions, i.e. very nearly $2c$.

We therefore accept the constancy of the velocity of light, as
measured by different observers, which is the cornerstone of
Einstein's Theory of Relativity, as an experimental fact. We
shall find that this has serious repercussions on our understanding of time.

Time dilation

It may seem, at first, as if the law regarding the constancy of
the speed of light, as determined by any observer irrespective
of his own motion, is interesting but of no great consequence.
Nothing could be further from the truth. Consider, for example,
the following situation. A space-ship has taken off from Earth,
has accelerated to some very high speed, has escaped the Earth's
gravitational field, and is now drifting through space with its
rockets turned off at a speed away from the Earth which, for
argument's sake, we might imagine to be half the speed of light
(i.e. $\frac{1}{2}c$).

Suppose now we wish to make contact with the occupants of this
space-ship. The only way in which we can do so is by means of
electromagnetic radiation. We can, for example, send a radio
signal out to the space-ship.

Such a radio signal leaves us with the velocity of light c.
We might at first imagine, therefore, that it would overhaul
the space-ship with a speed of only $\frac{1}{2}c$, but this is contrary to
the basic law concerning the velocity of light. The velocity

of light must always be the same relative to any observer irrespective of his own state of motion. *Thus the radio signal must still overhaul the space-ship with a velocity c relative to the space-ship.* In other words, if an occupant of the space-ship were to measure how rapidly this radio signal went past him he would obtain the same value c.

This seems impossible. How can a radio signal leave the Earth with a velocity c and yet still be measured to be travelling with a velocity c by occupants of the space-ship which is itself moving with a velocity $\frac{1}{2}c$ away from the Earth. *Einstein pointed out that the only way out of this dilemma is for an observer on Earth to assume that the rate of progress of time has slowed down on the space-ship due to its motion away from the Earth.* This is our only way out if we are to find the situation understandable. If a radio signal leaves us with velocity c and travels towards a space-ship moving away from us with velocity $\frac{1}{2}c$, we can only assume that it is catching up on a space-ship with a speed less than c. When we know from the basic law, however, that observers on the space-ship must measure the velocity of the signal relative to the ship also to be c, we must assume that time for them is passing more slowly than for us. For, of course, the velocity c which they measure is in so many km/sec in terms of *their* seconds, and if their seconds are appropriately longer than ours, they can still obtain the value c for a velocity which we think should be less than c.

Of course, everything is purely relative. If the occupants of the space-ship send a return radio signal to us they know the signal leaves them with the velocity c. At the same time, to them the Earth is moving away with velocity $\frac{1}{2}c$. Thus they would expect their signal to overhaul the Earth with a speed of only $\frac{1}{2}c$. The fact that they know, from the basic law of light, that the inhabitants of the Earth still measure this signal to be reaching them with the velocity c can only be understood by them if they assume that the progress of time on Earth is slower than on the space-ship. The fact that this conclusion, i.e. that each thinks that the other's time is slower than his own, is totally against all common sense merely shows that common sense is not a reliable guide in regions that are outside our common sense experience.

In the above example we assumed the space-ship to be moving away from the Earth with the velocity $\frac{1}{2}c$, but the argument is the same, whatever the relative velocity v between two observers and whether they are in space-ships or not. To see how this

relative motion affects the view that each has of the other's
measurement of time, we set up an imagined experiment.* We consider each of the two observers to have a long mirror which they
hold in such a way that the two mirrors remain parallel to each
other and 1 m apart throughout the motion (see Fig. 5.3). Each

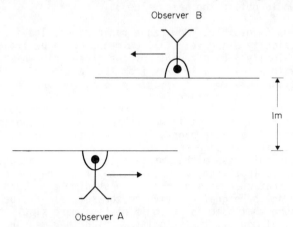

Fig. 5.3. An imagined experiment to compare
time between moving observers.

will take the view that he himself is at rest, while the other
moves with velocity v past him. Each now constructs a clock
for himself, the timing mechanism of which is based on the time
taken for a pulse of light to travel from himself to the other's
mirror and back. Let us consider the view point of observer A.
According to him, his light pulse will go from A to P and back
again, a distance of 2 m, in time $2/c$ (see Fig. 5.4), while that

_____

*This experiment can only be imagined and not actually performed,
since if the effects to be described are to be observed, the
two observers would have to travel at a speed near to that of
light. Imagined experiments are useful in stimulating thought,
but any conclusions reached must of course be checked by actual
experimentation. Einstein frequently used imagined experiments
in which railway trains moved at speeds near to that of light in
order to illustrate his theory, but the verification of the theory
rests on its agreement with actual experiments, such as the ones
described later in this chapter.

of B will have to go from B to Q in the time that B himself goes to D. To calculate this time we note that it is equal to either BQ/c or BD/v, depending on whether we consider the motion of the light pulse or the observer, so that

$$\frac{BQ}{c} = \frac{BD}{v} \qquad \text{or} \qquad BD = \frac{v}{c} \, BQ$$

We now apply Pythagoras' theorem to the triangle BDQ,

$$BQ^2 = BD^2 + DQ^2$$

Remembering that $BD = \frac{v}{c} \, BQ$ and $DQ = 1$, we have

$$BQ^2 = \frac{v^2}{c^2} \, BQ^2 + 1 \quad \text{or} \quad \left(1 - \frac{v^2}{c^2}\right) BQ^2 = 1.$$

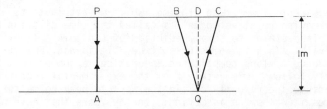

Fig. 5.4. Paths of light rays in the experiment of Fig. 5.3, as seen by observer A.

Finally, we use this expression to obtain the time taken by a light pulse for the path $BQ + QC$,

$$\frac{2BQ}{c} = \frac{2}{c\sqrt{1 - \dfrac{v^2}{c^2}}}$$

compared to the time taken by light for the journey $AP + PA$ which is

$$t = \frac{2}{c}$$

According to observer A, the clock of observer B therefore ticks slowly, by a factor $\sqrt{1 - \dfrac{v^2}{c^2}}$. However, since the situation between A and B is completely equivalent, we can go through exactly the same argument with A replaced by B, and we shall then come to the conclusion that according to observer B, the clock of observer A ticks more slowly. In general then, any observer will conclude that his clock ticks faster than the clocks held by any other observer who moves relative to him, and that the ratio of the rate of his clock to that of someone moving relative to him with velocity v is given by the quantity

$$\frac{1}{\sqrt{1 - \dfrac{v^2}{c^2}}} .$$

This apparent slower rate of time, which we attribute to any system moving relative to us, is called the time dilation effect. It is often stated to be a prediction of Einstein's Theory of Relativity. This is, of course, true. It is not, however, a consequence of long complicated theoretical argument, but quali-tatively follows immediately from the experimental fact that we must assume the velocity of light always to be c no matter who measures it or how he is moving. In the years that have passed since Einstein's theory there have been several experiments which indicate that the time dilation effect really occurs.

One such experiment concerns the muon, which is another of the unstable particles created in high energy nuclear reactions. It has a measured lifetime of only about two millionths of a second, after which it changes into an electron. Muons that are created by cosmic ray bombardment in the upper atmosphere at heights of about 30 km with speeds approaching that of light have been found to reach ground level, although even at the speed of light this distance takes one ten-thousandth of a second to travel. This is possible, because according to the observer on the ground the muon clock ticks much more slowly than his own. Using the above formula for time dilation, we find that in order to stretch the muon's life by the necessary factor of about 50, the speed of the muon must be within 1/10 per-cent of the velocity of light.

To see why we have not ourselves observed this time dilation effect before, let us look in more detail at the expression

$$\frac{1}{\sqrt{1 - \dfrac{v^2}{c^2}}}$$

which is plotted against v/c in Fig. 5.5. Clearly, as long as v is much smaller than c, the expression differs little from unity, and not until v is of the order 0.2 c, i.e. v = 60,000 km/sec which is a very large velocity indeed, can a departure from unity be detected on the graph. After that growth is increasingly rapid and, as v comes close to c, the expression tends to infinity. If we ever had an object moving with the velocity of light, then its clock would appear to us to go infinitely slowly, i.e. time for it would stand still. This is one indication that the speed of light must be greater than any other speed, and that in fact any massive object can never reach it, although as we have seen above, it can get very near to it.

Time dilation leads us, as an interesting sidelight, to the famous *clock paradox*. Suppose the space-ship of our example takes off from Earth and goes on a prolonged space voyage. As long as it is going away from us we must assume that its clocks are running slow compared to ours, and similarly on its return voyage the same is still true. Thus when the space-ship returns and lands we would have to expect that it has passed through less time than Earth has; the space-ship might have been away for 5 years according to our clocks, but to the space-ship's clocks and everybody in the space-ship, the time of the journey may have been considerably less - say 6 months. (We are assuming here that the space-ship was actually able to travel quite close to the speed of light.) Is such a thing theoretically possible?

The apparent paradox arises if one imagines one's self in turn on the space-ship. To the space traveller, the Earth is first moving away from him and therefore the time on Earth appears to be passing more slowly; after he turns around the Earth moves back towards him, but its time continues to pass more slowly than his. Would not the space traveller then find that the Earth had passed through less time than he had?

This is a question which we shall not go into in great detail here except to give the answer. It is indeed the first of the above cases which will apply, i.e. that overall the time passed through on the space-ship will be less than on Earth. This is the conclusion that one reaches as an Earth observer, noting

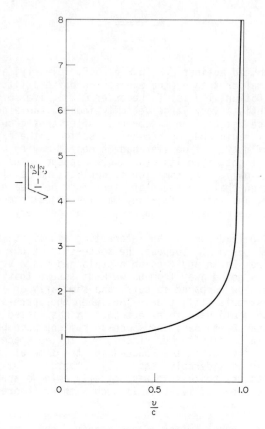

Fig. 5.5. The quantity $\dfrac{1}{\sqrt{1 - \dfrac{v^2}{c^2}}}$ gets

suddenly very large, as v
approaches c.

that the space-ship's time appears to be always running slow.
The reason why the second argument is not true is that there is,
in fact, an overall difference between the observers on Earth
and the people who went on the space voyage.

As long as the space-ship is drifting with uniform speed away
from the Earth, the two systems are completely equivalent, as
we noted earlier on, and each set of observers would reckon that
the time on the other system is running more slowly than their
own, and as long as the space-ship keeps drifting endlessly
through space each set of observers will always keep thinking
the same thing. However, in order to return to Earth the space-
ship must subject itself to accelerations of one sort or another,
in slowing down or turning around, or in any other manoeuvre.
Accelerations can be very obviously felt - particularly accelerations much in excess of that due to gravity on Earth - and thus
the space travellers do undergo very different experiences in
their journey through space and time from the people remaining
on Earth. A detailed study of this problem in relativity shows
that the results obtained by the observers on Earth from their
non-accelerated point of view, by which they conclude the space-
ship's time always runs more slowly, are correct. For the
corresponding analysis for a traveller on the space-ship, account
must be taken of all periods of acceleration in a different way
from the simple relativity arguments of this section, which
apply only to motions with constant velocity, i.e. with no acceleration. When a full theoretical analysis is made, including
the accelerated periods for the space traveller, the two results
are in agreement, and so indeed, the space traveller does find
himself younger on his return than his companions on Earth.

Recently, an experiment has been performed by the American physicist Hafele, which aimed at a direct verification of the paradox.
He took four atomic clocks round the Earth both eastwards and
westwards, using commercial jet flights, and compared them with
a stationary clock, both before and after the flight. Because
of the rotation of the Earth, the effect is not the same in the
two directions, and Hafele predicted that the moving clocks
would gain 275 nanoseconds in the westward flight and lose 40
nanoseconds in the eastward flight (1 nanosecond = 1 thousand
millionth part of a second). The actual results were 160 and
50 nanoseconds respectively, close enough to indicate a verification of the paradox. (For further details, see G. Wick,
The clock paradox resolved, *New Scientist,* 3 February 1972,
p. 261.)

Simultaneity revisited

We next return to another consideration of Chapter 4. There we
concluded that we could not synchronize distant clocks, because
we could not be sure that light velocities in opposite directions

were the same. But now we know that they are, and so it would
appear that we can after all synchronize our clocks at *A* and *B*
by sending out a light signal in opposite directions from *C*,
the mid-point of *AB*. To show that the matter is more complicated, Einstein imagined a railway carriage passing through a
station* (see Fig. 5.6). The guard at *C* in the exact middle of
the carriage presses a button the moment he passes the station
master *D* on the platform. This button releases light signals
in opposite directions which actuate a mechanism that opens the
doors *A* and *B* at opposite ends of the carriage. As *C* is exactly
in the middle of the carriage, the signals travelling with the
same velocity will reach the two ends of the carriage at the
same time and the guard will see the doors open simultaneously.
Not so the station master at *D*. He too will observe the two
light signals to have the same velocity, according to Einstein's
postulate, but since the carriage is travelling in the direction
AB, the rear end of *A* will move to meet its signal, while the
front end at *B* moves away from its signal. For that reason the
station master will conclude that the door at *A* opens before the
door at *B*.** Similarly, an observer in another train that is
overtaking the first train will conclude that the door at *B*
opened before the door at *A*.

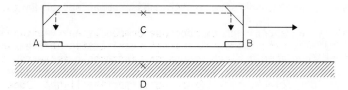

Fig. 5.6.   Events that are simultaneous for *C*
are not simultaneous for *D*.

We conclude that it follows from the postulate of the constancy
of the velocity of light that two events that appear simultaneous
to one observer may not do so to another. Further, while the
station master sees door *A* opening before door *B*, a passenger

---

*Here we have another of Einstein's imagined experiments.
**It should be noted that this is quite separate from the fact
  that in addition the door at *A* will be nearer to him when it
  opens, than that at *B*, so that the observation of the door at
  A opening will take less time to reach him.

travelling in a second train in the opposite direction to the
first train, will see door B open before door A. Thus the time
ordering of events can be reversed for different observers, i.e.
while one observer could conclude of two events, which we may
call A and B, that A preceded B, another observer could conclude
that B preceded A. This is certainly surprising, but it need
not cause us serious concern unless one of the observers should
conclude that there is a causal connection between A and B. In
that case we clearly would be in great difficulty, for if A
caused B, it is obviously necessary for A to precede B for all
observers. In the case under consideration this is not so,
since the opening of both doors is due to the pressing of a
button at C, and it is easily seen that it would actually be
impossible for a light signal to pass from A to B in the time
available between the opening of the doors. Fortunately, it can
be shown that the kind of reversal of A and B postulated above
can occur only when there is no causal relationship between them
and that when the time interval between two events is sufficiently long for a light signal to be able to pass between them,
so that a cause and effect relationship becomes possible, the
ordering of the events is the same for all observers. It is
only when events are clearly independent of each other that their
time ordering can change.

Conclusion

Our brief excursion into the Theory of Relativity has led us to
realize that the nature of time is a great deal more subtle than
could be appreciated from our sense impressions alone. We have
had to abandon some concepts which were so firmly held that they
had probably never been questioned before. These included the
concept of a time flowing constantly for all observers, the concept of the simultaneity of events and even that of the ordering
of events in time. At the same time, it is reassuring that we
have not had to abandon the principle of causality, which we
could hardly have done and still kept to the fundamental logic
of science.

Basically, it all came out of the time dilation effect which
said that a clock moving relative to an observer had its time
scale lengthened by a factor

$$\frac{1}{\sqrt{1 - \frac{v^2}{c^2}}}$$

compared to the observer's own stationary clock. But let us
not forget that, even for an astronaut, v is only about a ten
thousandth part of c, so that an astronaut's clock, as seen by
us on earth, is slowed down by a factor

$$1.000000005.$$

Finally, the fact that the velocity of light is a universally
constant quantity, means that we can measure distances by
measuring the time taken by light to travel over them. Methods
of this kind, such as radar, are actually the most accurate
methods of distance measurement known, and they illustrate the
conclusion, originally reached by Einstein, that space and time
are closely related.

Chapter 6

THE DIRECTION OF TIME

The flow of time

When, in Chapter 3, we looked at our intuitive experience of
time from the point of view of specifying the properties that a
time-measuring device must possess, we concluded that one of the
most important of these was the way that time always moved in
the same direction. In this it was essentially different from
space - the distinction between past and future was clearly much
more fundamental than that between left and right or up and down,
it was not possible to visit points in time at will in a way
that it was, at least in principle, possible to visit points in
space and so on.

When we came to the construction of clocks, we had to conclude,
however, that it was not the direction of flow that was important,
but the rate of flow. Further, it turned out that this was
measured much more accurately by devices that incorporated a
rhythmic change than those that incorporated a flow change.

Obviously, our experience of time involves both permanent con-
tinual changes and regular rhythmic changes. The former are a
good measure of the direction of flow, but they give little idea
of its rate; the latter are a good measure of the rate of flow,
but they give no idea whatever of its direction. We therefore
have the paradoxical situation that none of our modern time-
measuring devices measure the most fundamental and incisive property of time - its direction. The reason for this is, of course,
that we do not need a measuring device to do this; our intuitive
knowledge of time is quite adequate for it in general. Why this
is so will be the main line of enquiry in this chapter.

Let us analyse this intuitive knowledge further. If we see two
photographs of a cup, one held in a hand and the other with it
smashed on the floor, and we are told that these are pictures
of the same cup taken 2 seconds apart, then we know that the one
with the cup on the floor must have been the later one. Similarly, we know that when we drop a hot iron bar into cold water,

we finish up with a warm bar in warm water, but that if we start
with a warm bar in warm water. it will never turn into a hot bar
in cold water, far less jump out of the water into our hand.
There exists a film taken backwards of a man eating a banana.
Each time he takes a bite, more banana comes out of his mouth,
until eventually he holds a whole banana in his hand which he
proceeds to wrap up carefully in its skin. It is very funny to
watch but it is obviously impossible. Why?

Order and disorder

If we try to analyse what is common to the situations described
above, we find that in some sense the systems observed had a
greater degree of order at the beginning than at the end. A cup
is a more ordered arrangement of the material of which it is
made than a heap of broken sherds; a banana, after it has been
chewed, is a less orderly arrangement than before. The situation
is a little less obvious with the iron bar and water, until we
recall that heat is due to the motion of molecules and that the
hotter a body, the faster its molecules move. Thus with the hot
bar in cold water the majority of the fast moving molecules are
in the bar and the slow moving ones are mostly in the water,
while there is no such separation in the case of the warm bar in
warm water. This conclusion, that things not only change, but
change in such a way that they show less order, is indeed one
that seems to be based on very general experience. "Change and
decay in all around I see", as the hymn says. Not only change,
but also decay, and decay, whether it is the crumbling of a
statue over the centuries in wind and rain, the floor of a
forest that is made up of leaves that have fallen from the trees,
or what is left of poor Yorick after he has been in the ground
for eight years, all leads to a lessening of order.

Of course, there are situations where this is not so. The
statue, the tree and poor Yorick at one stage all showed very
substantial increases in order over the previous order that
existed in the materials from which they were made. Even in the
purely inanimate world there are instances like the growth of
crystals from solution, where there is clearly an increase in
order. The question arises, whether such increases in order
are purely localized and are in fact paid for by a greater increase in disorder elsewhere. Thus the part of the food which
an animal uses in maintaining its growth finishes up in a more
ordered state inside its body, while that which it eventually
excretes as waste finishes up in a less ordered state. To esta-
blish whether there is a net increase in order or disorder, we

would first of all have to devise a numerical measure of order
and then an appropriate measuring device. We shall return very
briefly to this later, but in the meantime tentatively postulate
that *in any isolated system disorder increases with time.* This
is certainly in line with our experience, to the extent that
simple systems for which we found the statement to be true, e.g.
the iron bar in the water, can effectively be isolated, while
systems in which order appears to increase are manifestly joined
to other systems from which they cannot be isolated, e.g. the
tree to the air and soil from which it obtains its food and
eventually to the Sun. If these are then included in a larger
system, we face the problem of the balance between local increases
of order and disorder within the enlarged system.

Order and probability

How can we quantify order? If we throw a dice six times, and
the first throw showed a one, the second a two and so on up to
six, then we would feel that there was a very high degree of
order of throws here. If we threw it again six times, and again
obtained a one, two and so on up to six, but not in the order
in which we threw the dice, then we would judge that, while
there was still a lot of order around, it was less than before.
Now of all the ways, and there are 6 x 6 x 6 x 6 x 6 x 6 = 46,656
of them, that we can throw one dice six times - only one will
give the first result, while there are 720 ways that lead to
the second result. (The first throw can be in any one of six
ways, the second in five ways and so on, leading to a total of
6 x 5 x 4 x 3 x 2 x 1 = 720 ways.) Obviously, therefore, the
probability of obtaining the first result is much smaller than
that of obtaining the second result.

We are led here to associate an increase in disorder with a
higher degree of probability of achieving it, and we may be
reminded of the statement that a monkey hitting the keys of a
typewriter at random will, given enough time, type out the plays
of Shakespeare. He will of course need a lot of time, because
the probability of achieving the result is small, and this is
associated with the fact that the plays, which consist of a
meaningful assembly of the letters of the alphabet, exhibit a
high degree of order of these letters.

It is intuitively obvious that the more individual items are
contained in an assembly of such items, the less likely is it
that they will show any significant amount of order. Now the
number of letters in the plays of Shakespeare is about 10,000,000,

which is tiny compared to the number of molecules in even a
single drop of water, which is about

$$3,000,000,000,000,000,000,000.$$

Such a number is unimaginably large, but to give some feeling
for it let us look at it in terms of the old story of the man
who invented the game of chess. When asked for a reward, he
requested that he be paid in wheat, one grain on the first
square, two on the second, doubling the quantity each time until
the last square. This would result in the unbelievable number
of about

$$60,000,000,000,000,000,000$$

grains on the board, a total that could not be carried by all
of today's shipping fleets. If, on the other hand, he had
asked for one molecule on the first square, two on the second
and so on, then the total would not have been enough for a single
grain of wheat.

We may now begin to see a reason for the postulate that disorder
in a system increases with time. The systems with which we are
concerned are made up of an unimaginably large number of molecules and their behaviour is due to the averaged behaviour of
all these molecules. Under these circumstances, states of increasing disorder are highly probable, while significant ordering becomes improbable to a degree that makes it effectively
impossible. If we consider, for instance, the molecules of gas
in a vessel and ask for the probability of all of them at a
given time being in one half of the vessel, leaving the other
half empty, then because of the very large number of molecules
in even a small volume of gas, the probability of this happening is exceedingly small, and the degree of order, if it did
happen, correspondingly high. The astronomer Eddington once
expressed this by saying that the likelihood of it happening
was considerably less than that of an army of monkeys typing
out all the books in the British Museum, by striking the typewriter keys at random. We therefore have a mechanism that
explains the existence of our postulate that disorder in a closed
system increases with time, in terms of probabilities. Most
physical laws are not in so fortunate a position, in that there
is always the possibility that sooner or later an event will
turn up that will lead to a contradiction with any particular
law. With our postulate we actually expect this to happen, but
we also know how small the probability is of it actually happening. This greatly strengthens our belief in the validity of the
postulate.

Relation to thermodynamics

The problems of order and disorder were first investigated in
connection with the kinetic theory of matter, which treats matter
as being made up of molecules in motion. In this theory, as has
already been indicated, the increase in energy due to heat being
supplied to a quantity of matter is interpreted as resulting in
an increase in the kinetic energy, that is energy of motion, of
the individual molecules, in other words, the average velocity
and hence the average kinetic energy of the molecules in a hot
material is greater than the corresponding quantities in the
same material, when cold. The science which deals with this
relationship of heat and motion is called Thermodynamics, and
it is also concerned with such practical matters as the way heat
energy is turned into mechanical energy in, say, an internal
combustion engine. In fact the First Law of Thermodynamics
states that the mechanical energy produced in such an engine
can never exceed the amount of heat energy put in. This law,
which is a special case of the Law of Conservation of Energy,
prevents us from making perpetual motion machines, which would
produce mechanical energy in excess of the heat energy supplied.
Another way of phrasing the First Law of Thermodynamics is
therefore that "You can't get something for nothing."

When heat engines were further investigated, it turned out that
the First Law of Thermodynamics was not enough to explain their
action. According to this law, it would be perfectly possible
to have an engine which, starting at the temperature of its
surroundings, would get colder and colder, with the heat energy
that was liberated in this way being turned into mechanical
energy. Such a situation, in which one starts with everything
in a system at the same temperature and finishes with part of
the system hot and another part cold, exists of course in a
refrigerator, but there it is achieved through the application
of mechanical, electrical or even heat energy. The mechanical
energy which we can then derive from the hot part of the system
is always less than that which we put into it in the first place,
so that the system does not work as a heat engine. On the other
hand, the reverse process, by which a system with parts initially
at different temperatures finishes up all at the same temperature, is perfectly possible and does not require an energy
supply.

The fact that it is necessary to supply energy in order to produce a difference in temperature within a system, where initially
there is none, cannot be a consequence of the First Law, since,

if it were, the reverse process would also be forbidden. We
note, however, that the process involves a change in the total
order of the system since, as we explained before, order is
increased when a system separates into a hot and cold part.
That such a change is not permitted follows from the postulate,
that we have been discussing for most of this chapter, and to
which we can now at last give its official title, the Second
Law of Thermodynamics. When C. P. Snow, in his famous lecture
on "The Two Cultures", remarked that for scientists to be
unfamiliar with Shakespeare was no worse than for humanists to
be unfamiliar with the Second Law of Thermodynamics, he was
perfectly right, for this law, probably more than any other,
governs not only our lives, but the behaviour of the whole
Universe. We therefore restate it as follows:

> "Changes in any isolated system always occur in
> a way such that the order of the system
> decreases."

In other words, "It takes work to prevent chaos."

It may be a little worrying that we have introduced the concept
of order into this law, without having properly quantified it.
If you are not worried about this, you may skip the rest of this
section, but if you are, then you may like to know how scientists
got over this problem. It is in fact possible to work out the
probability that a system changes from one state to another by
counting all the possible ways that the second state can happen,
as was indicated above in the case of dice. Let us call this
probability P. Now to obtain the probability of two successive
changes occurring, we have to multiply together the probabilities
for each change to occur separately. This is illustrated by
for instance noticing that the probability of throwing a six
with one dice is 1/6, but the probability of throwing two successive sixes is 1/36 = 1/6 x 1/6. On the whole, we find it
easier to handle quantities that add than those that multiply,
when they describe successive events. We now remember that if
two numbers multiply, then their logarithms add. We therefore
define a new measure of disorder, which is called entropy, from
the Greek word *trope*, meaning change, and given the letter S,
through the equation

$$S = k \log P$$

where k is a number which we do not need to specify further for
our purposes. Then, if the probabilities of two events are P_1 and

P_2, the probability of their happening in succession is $P = P_1P_2$ and the corresponding entropy is

$$S = k \log P_1P_2 = k(\log P_1 + \log P_2) = S_1 + S_2$$

which is equal to the entropy associated with each event separately.

It is possible to express the increase in entropy as a system changes from one state to another in terms of the amount of energy supplied to the system in the change, Q, and the temperature of the system, T. (The latter must be measured from the absolute zero of temperature, which is minus 273°C.) It can be shown that in such a change, the entropy increases by Q/T.* That this is qualitatively reasonable can be seen by thinking of a gas which consists of molecules in random motion. Heating the gas, i.e. adding energy, produces more motion, more collisions, more disorder. But in proportion this effect will be greater if the previous disorder was less, which will be the case at lower temperatures. Hence the disorder and the entropy increase more for a given Q and lower T, just as they increase more at a given T for greater Q.

If we now think back to our system of the iron bar and the water, then we had a situation there where heat flowed from the one to the other. Let us assume that the bar was at absolute temperature T_1 and the water at T_2 and that the heat transferred was Q. We now calculate the change in entropy, which is the sum of the ratios of the heat gained to the absolute temperature for all parts of the system. It is equal to

$$\frac{-Q}{T_1} + \frac{Q}{T_2}$$

As T_1 is greater than T_2, this quantity is clearly positive, which means that heat flows so that the entropy increases. This is a simple illustration of a very general result that in any change in an isolated system the entropy always increases. This is another and more quantitative form of the Second Law. However, from our point of view, the relation to the direction of

---

*For those who know something about thermodynamics, it should be added that changes in the system due to the increase in energy must take place infinitely slowly, so that the system finishes up in equilibrium.

the flow of time, the expression of the law in terms of the
increase of disorder, is very much more helpful.

How universal is the Second Law?

Let us return to a consideration of time. What we have established is an association between the direction of the flow of
time and the increase in disorder of isolated systems. To the
extent that no system can be fully isolated from the rest of
the Universe, it may be conjectured that the increase in disorder is one of the Universe as a whole. Eventually, such a
universe would be in complete disorder, which means that it
would be at a constant temperature throughout. In such a universe nothing could happen and this has been referred to as the
ultimate "heat death" of the Universe. It is of course dangerous to extrapolate so far beyond our very limited experience,
either by postulating the increase in disorder for the Universe
as a whole, or by taking the matter to its apparent logical conclusion and postulating an ultimate heat death. It would be
wise at present to submit the case to a Scottish jury, which in
addition to the verdicts of guilty and not guilty is able to
pronounce a verdict of not proven. There is, incidentally, a
possible fallacy in the "heat death" argument and that is that
the Second Law only applies to an isolated system, i.e. to
changes in entropy due to the interaction of parts of a system
which itself is isolated from the rest of the Universe. As the
Universe encompasses all, it may therefore be meaningless to
call it an isolated system.

There remains a fascinating possibility of linking cosmological
processes to the direction of flow of time. All the evidence
available suggests that our Universe is expanding, a point to
be taken up in Chapter 7, so that here we have a unidirectional
phenomenon on the cosmic scale. It is tempting to link this to
the other unidirectional phenomenon that is so all pervading,
time, but at present this is as far as scientists appear to have
got. (See Bondi's remarks in *The Nature of Time*, ed. T. Gold,
Cornell University Press, 1967, p. 4.)

A verdict of not proven must at present be our attitude to the
working of the Second Law in biological processes, since no
methods exist for these which will determine exactly whether in
such processes there is a net increase of disorder, i.e. entropy.
However, to the extent that we know of no biological violation
of the Second Law, nor of any biological violation of physical
laws in general, it seems reasonable to include biological

processes within the realm of the Second Law. It would be
undesirable to make any special explanations for biology when
there is no clear need for it.

Microscopic systems

So far we have concerned ourselves with so-called *macroscopic
systems*, i.e. systems consisting of very many particles in which
we are concerned with overall properties, such as temperature,
pressure, etc. However, these are made up of a very large number of *microscopic systems*, i.e. individual particles in interaction with other individual particles. In principle, it should
be possible to derive the laws governing macroscopic systems
from those governing microscopic systems. Now the latter are
very well known; they are the laws of motion, together with the
laws of gravitational, electromagnetic and possibly nuclear
interactions, and it is a curious fact that, with one possible
exception to be mentioned below, none of these laws distinguish
between time going forward and backward. What we mean by this
is that if we have, say, an atomic collision process, in which
two atoms approach each other under their mutual interaction and
then separate again, then the reverse process is just as possible
(see Fig. 6.1) and this is obtained from the relevant equations
by replacing the time t by minus t everywhere. (This is called
microscopic time reversal.) Such a collision can be simulated

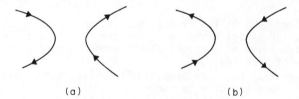

(a)   (b)

Fig. 6.1.   If (a) is possible, then (b) is
            possible.

using dry ice magnetic pucks, which are almost frictionless,
and photographing them at regular intervals by means of strobo-
scopic light. It is clearly extremely difficult from the final
photograph (see Fig. 6.2) to tell the direction in which the
pucks moved, but because of the very slight friction, which
slows the pucks up, it is in fact just possible to do so. The
pucks are of course macroscopic systems, for which time cannot
be reversed, and the friction effect is a manifestation of the

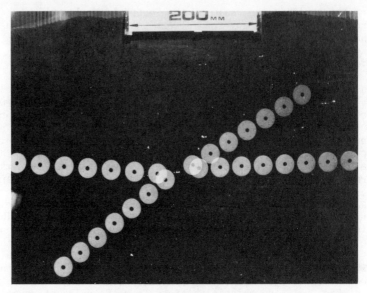

Fig. 6.2. Collision of almost frictionless
magnetic pucks. (Courtesy of
Dr. J. M. Kilty).

Second Law. But on the atomic scale, there is no friction, and
the question arises as to how time-irreversible macroscopic processes can arise as a result of microscopic ones, that have the
property of time reversal.

The answer that has been proposed is subtle, and we shall illustrate it in terms of a gas that at one stage occupies half the
available volume of a vessel and at another all of it. This
can in practice be achieved by having a membrane across the
vessel, evacuating the gas from one half of the vessel and then
removing the membrane. The gas will then spread throughout the
vessel. Now let us assume that at some point in time the molecules are in certain positions with certain velocities, and that
at some later time they are in the same position again, and also
with the same velocities, except that now all the velocities are
in the opposite direction to what they were before. If this
ever happened, then the molecules would from that moment on retrace their previous paths and the gas would eventually return
to its initial state, filling one half of the vessel. Clearly,
in principle all this could happen, it just happens to be fantas-

tically unlikely, because of the very large number of molecules
involved. It therefore becomes a matter of probability and not
of certainty that macroscopic processes in general go from order
to disorder, which is of course what we said before.
Another objection to the statistical explanation of the concept
of time is the so-called periodicity objection. However small
the probability of a particular ordered state may be, given a
sufficiently long time, it may be expected to turn up, i.e.
sooner or later these monkeys *will* type out the plays of
Shakespeare. This means that at some stage, statistical fluctuations may lead to an increase of order. This objection too can
be met by saying that the Second Law does not postulate that an
increase in order is impossible, but merely that it is highly
improbable. Nevertheless, the two objections that we have listed
are sufficiently real for scientists to have cast serious doubt
on the validity of the statistical interpretation.

Breakdown of microscopic time reversal

We conclude this chapter with a brief reference to the possibility of a breakdown in microscopic time reversal. For a long
time, it was assumed that all processes between elementary
particles were symmetric in both space and time. Then in 1957
an experiment was performed which showed that certain elementary
particles could distinguish between left and right, thus intro-
ducing a degree of asymmetry into space. More recently there
has been similar evidence for an asymmetry in time. In essence
this would imply that there are processes between elementary
particles for which the simple scheme of Fig. 6.1 did not apply.
Such a breakdown in microscopic time reversal would have certain
implications regarding the structure of the laws of nature,
which are of considerable interest to physicists. It would not,
however, of itself lead to the kind of logical absurdities such
as reversal of cause and effect that macroscopic time reversi-
bility would lead to.

Chapter 7

TIME AND THE UNIVERSE

The Universe

Up to now our investigations have been essentially earth-bound,
and we shall see in the present chapter that we shall obtain
further illumination regarding the nature of time, by studying
the relationship of time to the Universe as a whole.

We are but living beings on one single planet in the entire
Universe. The Universe is not static. Innumerable changes are
continuously occurring within it, changes which are completely
unaffected by our existence. New suns are in the process of
formation, old suns are "dying", new galaxies are being formed —
things are happening with the passage of time in the Universe.

From our previous discussion it is clear that in any discussion
of the Universe in general the concept of space - that is, the
distance between objects in the Universe - is far from sufficient. One must also consider the behaviour of the Universe as
regards time. It will be our task in the present chapter to
give an elementary introduction to the concept of time in the
Universe, and to this end we devote this section to building up
a picture of how the Universe appears to us and of where we, and
our solar system, stand in relation to it.

The things that seem normal to us and to which we are accustomed
are things on Earth because, after all, that is where we live;
but while what happens to us on Earth is certainly important,
we are in fact a tiny speck in the Universe. For instance, our
Earth is only a satellite of the Sun, an object, which both for
size and temperature is beyond our comprehension. Its diameter
is more than three times the distance from the Earth to the Moon
and the temperature in its interior is measured in millions of
degrees. It produces its heat and light by nuclear reactions
similar to those in a hydrogen bomb, deep in its central region.
This has been going on for several thousand million years. Yet
there is still so much matter in the Sun that it will continue
exploding and producing heat for at least several thousand

million more years. This great object has a number of satellites - planets - revolving around it, one of which is Earth.

The distance of the Earth from the Sun is approximately 150
million kilometres, while that of the furthest planet, Pluto,
is nearly 6000 million kilometres. These distances may seem
large, beyond our imagination, but the distance to even the
nearest fixed star, the brighter of the two pointers of the
Southern Cross, is very much larger, about 40 million million
kilometres. Distances as large as this are usually measured by
the time that light takes to travel them, and since light travels
at the rate of 300,000 km per second, one *light year* equals
about 10 million million kilometres. We then find that the distance from the Earth to the Moon is about 1 light second, to the
Sun about 8 light minutes, to Pluto over 5 light hours, and to
the nearest fixed star about 4 light years.

But even distant stars which we can see with the naked eye when
we look at the sky on a clear night are still nowhere near the
limits of the Universe. When scientists look at the heavens
through large telescopes they see that our Sun is but one of a
large group of suns which together we call a *galaxy*. There are
in fact something like 100,000 suns in this group or cluster of
suns which is our galaxy. We can also tell that all these suns
in our galaxy are formed into a pattern so that the overall
shape of the galaxy is flat but with a bulging centre - something like a fried egg hanging in space. So large is our group
of suns - our galaxy - that it takes light about a hundred thousand years to go right across it, and even to go straight through
the narrowest parts it takes light ten thousand years.

With large telescopes, however, it is possible to look much
further than this. Astronomers can look beyond our galaxy and
find that there is mostly empty space for millions of light
years. Eventually, however, the big telescopes show up in the
distance another cluster of stars somewhat like our own galaxy;
and then after another few million light years of empty space
there is another cluster of stars, and at greater distances
again, yet another, and again another, and so on. So the Universe goes on as far as we can see - which with today's largest
of telescopes is a few thousand million light years or several
thousand million million million kilometres. *It is an interesting thought that when we look at such a star cluster we are
seeing it not as it is now, but as it was when it emitted the
light several thousand million years ago. Looking a long way
away is also looking back in time.*

Each of the galaxies in the Universe contains from 100 million
to 100,000 million stars. Large numbers of these stars or suns
probably have planets revolving around them; and on each of
these planets there is some chance of life existing. What an
insignificant fraction of all the life in the Universe we may
form! It has been estimated that, even within our own galaxy,
which is itself such a speck in the Universe, there are possibly
50,000 million suns with planets revolving around them.

Time scale of the solar system

Within the scope of this book it is not appropriate that we
should go into great detail about modern theories of the formation of our solar system. Suffice to say that it is considered
to have formed from a large nebulous cloud of slowly rotating
gas which has contracted and condensed into the solar system
over millions of years.

The central part of the solar system is the Sun itself and it
is almost certain that many other stars of our galaxy also have
planetary systems which have been formed by the same type of
process. When gases contract to form a star, the young star
consists mainly of fundamental atomic particles - that is, protons and electrons and a relatively small number of heavier
atomic nuclei. As the Sun contracts under its own gravitational
forces it "heats up" and so violent are collisions between the
atomic particles or nuclei that nuclear reactions can take place
with release of energy - the source of all the radiation from
the Sun.

With our present understanding of nuclear processes it is possible to calculate what we believe must be going on within stars,
and it is found that each star must have a finite life. As the
original light elements such as hydrogen are used up in the
nuclear reactions and more and more heavier elements are manu-
factured, the star may even blow apart eventually. This occurs
when the amount of energy being released in the nuclear reactions
becomes so great that the internal gravitational forces can no
longer hold the star together.

Such an occurrence was actually observed from Earth in the year
1054, when a certain star suddenly began to glow with amazing
brilliance. Night after night it grew brighter and to those who
watched it, it appeared to be expanding. Eventually it started
to dim and gradually disappeared from view. A star had "died".

The description of this event has been found in Chinese writings.
Large telescopes, pointed to the spot where the star appeared,
show what remains; it is called the "Crab Nebula" and represents
the remnants of the star distributed in a great gaseous cloud
and still expanding outwards rapidly. Such an explosion of a
star - called a *super nova* - occurs in our galaxy about once
every fifty years on the average. After such an explosion,
atomic nuclei of the various elements which were formed in the
star are scattered across millions of miles of space. This may
well be the fate of our own Sun eventually. It has used up
about half of its original supply of hydrogen, but this has
already taken several thousand million years. Our Sun is in the
equilibrium stage, where it still has ample hydrogen left and is
radiating energy into space at the same rate as it is being
released within it; thus the Sun will probably last for a few
thousand million more years.

We thus see that from observations of the Sun, and from calculations regarding the nuclear reactions proceeding within it, we
estimate that our solar system must have an age of several thousand million years.

In the above we have discussed the age of our Sun and our solar
system. On the basis of observation and calculation the natural
life span of a typical sun would appear to be of the order of
10 thousand million years.

But just as the average life span of a typical human being says
nothing about the age of mankind in general, so the average life
span of a sun says nothing about the age of the Universe.
Although each person today may have an average life expectancy
of a little over 70 years, this has no bearing on the time over
which man has inhabited the Earth. However, before tackling the
problem of the age of the Universe, we have to learn more about
some of its other properties.

Olbers's paradox

In 1826 a German astronomer, Heinrich Olbers, published a remarkable paper in which he arrived at a famous paradox regarding the
Universe. Although it was not realized at the time the most
plausible solution to Olbers's paradox would, over a hundred
years ago, already have led scientists to the conclusion that
the Universe is expanding.

Olbers performed an extremely simple calculation. We all know
that the sky is black, or at least extremely dark, at night

except for the close visible stars of our own galaxy. Olbers
set himself the task to estimate the total amount of light which
should reach us from all the stars and distant galaxies.

The calculation depended on the simple assumption that the distribution of galaxies throughout the Universe is on the average
uniform, that is, the average density of galaxies in the Universe
is the same everywhere. This means that we assume that any
large volume - say a sphere of radius 100 million light years -
contains approximately the same number of galaxies.

Although galaxies differ one from another, we can, for simpli-
city, assume that the Universe is made up of "typical" galaxies,
all of which emit the same amount of radiation. Now the amount
of energy which we receive from any particular galaxy depends
on the distance between it and us. As will be apparent from
Fig. 7.1, the light from any source of light, say an electric
bulb, that falls on a square of side 1 cm, a distance 1 m away,
will cover a square of side 2 cm, that is an area four times as
large, at a distance 2 m away. The illumination that falls on
a square of given size is therefore four times less at twice
the distance, and quite generally it decreases in inverse proportion to the square of the distance. Exactly the same law
will hold for the energy which we receive from a galaxy. On the
other hand, if there is a certain number of galaxies, say one
million light years from us, then the same argument shows that
there will be four times as many on average that are two million
light years from us. Hence the total energy reaching us from
all the galaxies one million light years from us is the same as

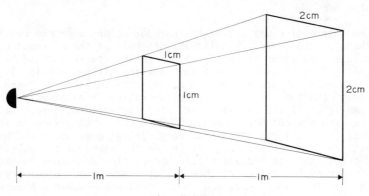

Fig. 7.1.

the total energy reaching us from all the galaxies that are two
million light years from us, and by the same reasoning the same
as the total energy from all the galaxies that are at any given
distance from us, however large the distance. There is therefore no diminution of the total energy coming to us from galaxies
with increasing distance, and if the Universe is infinite in
extent, then the total energy reaching us from all the galaxies
in it must be infinite too.

Naturally a correction has to be made to this first result
because the light from very distant galaxies has some chance of
being blocked or intercepted by closer galaxies. When this
correction is applied, the answer obtained is still very surprising. It is that the intensity of radiation at any point in
the Universe must be precisely the same as on the surface of a
typical star or sun, which is about 6000°C.

This is obviously not the case, since the surface of our own
Earth is not at such a high temperature or anywhere near it.
The result is therefore incorrect. Yet it is not that Olbers
made an absurd calculation, for the assumption of a uniform
Universe was a very straightforward one and the result followed
simply and logically from this. Olbers himself tried to resolve
the paradox by postulating that there must be some form of con-
tinuous gas between galaxies which gradually absorb radiation
travelling over very long distances and which could not be
detected by astronomical observations. This attempted explanation is not actually tenable, as even if this gas existed it
would reach an equilibrium temperature after which it would
radiate as much energy as it received and no longer act as an
absorber.

One interesting way out of the paradox occurs when one realizes
that, on the Olbers calculation, one-half of the enormous calculated radiation intensity at the Earth's surface would be
coming from galaxies more than a million million million light
years away. Suppose, therefore, that the Universe is "very
young" and that it only started "operating" a relatively short
time ago. If the galaxies were distributed through space as
they are now but were somehow "turned on" at some time in the
past, then the radiation from very distant ones would not have
yet had time to reach us. If indeed the stars of the galaxies
did not start to radiate until some particular moment in the
history of the Universe, then the darkness of our night sky tells
us that this would have had to be somewhere between a hundred
million and a million million years ago.

For in this case any galaxies further away than this would have
radiation on its way to us but not as yet having reached us.
On this explanation the night sky would gradually, over the hundreds of millions of years, be becoming brighter and brighter
as more and more radiation arrived and eventually the Earth
would "fry".

It is to be noticed, however, that the time of "switching on"
of the stars of the Universe in this explanation would be
remarkably "recent" in history. The lower estimate of a hundred
million years can be ruled out simply on the basis that we know
the Earth is older than this, although the upper limit of a
million million years would be permissible on these grounds.

We now know, however, that no such *ad hoc* switching on process
for the Universe is needed to resolve Olbers's paradox. One
solution which was not considered by Olbers was discovered by
direct astronomical observation in the present century: *the
Universe is expanding*.

When spectroscopic photographs are taken of the light from distant galaxies, it is observed that the characteristic spectral
lines from a known element have lower frequencies than on Earth,
so that they are shifted for instance from the visible region
into the infrared region. This effect, known as the *Doppler
effect*, always results when the source of a wave is moving away
from the observer, and is well-known in acoustics. (We referred
to it in Fig. 5.2.) Thus the sound of an aeroplane drops in
pitch as the plane passes overhead and recedes into the distance.
The similar effect observed in the light of galaxies indicates
that they too move away from us.

Now it can be shown that the energy carried by radiation, such
as light, is proportional to its frequency. Further, the astronomical observation shows that the further a galaxy is away from
us, the faster away it is moving. When one goes to the extreme
distant galaxies such as enter in the Olbers calculation, one
reaches the situation when wavelengths normally in the visible
region reach us way down in, say, the infrared or even radio
region. Thus it is quite understandable that the night sky is
dark simply because of the Doppler change in characteristics of
the radiation reaching us from a distant spherical shell independent of the radius of that shell because the larger the radius
the faster are the galaxies within it moving away and the less
the energy reaching us.

It is an extremely interesting concept, however, that our night
sky is dark, and that the surface of the Earth is not at some-

thing like 6000°C, and in fact that the Earth and the life on
it can exist only because distant galaxies are behaving in a
certain way. In this sense our existence is dependent on what
is happening to galaxies even more than a million million million
light years away.

The age of the Universe

By now, detailed observations of the "red shift" - that is, the
Doppler effect - of a large number of distant galaxies have been
performed and a simple law for the expansion of the Universe
deduced.

This simple law is based on what is called the *cosmological
principle* in which it is assumed that the expansion of the Universe is going on everywhere and that it would appear to be expanding in the same way from each point in the Universe. This
means that not only do we on Earth see all distant galaxies
moving away from us, but supposing we were on a planet in some
distant galaxy say a million million light years away, then the
cosmological principle says that we would still see all other
galaxies moving away from us according to the same law as we
would derive here on Earth.

This is not a very difficult concept to appreciate as can be
seen by a one-dimensional analogy. Suppose we consider a piece
of elastic which is being stretched at a uniform rate. If initially there were marks along the elastic say 1 inch apart
(imagine the elastic to be initially several feet long) then as
the elastic is stretched so each mark gets further away from its
neighbours. Any "observer" situated at one of the marks would
see the nearest marks on either side getting further away, the
next nearest marks on either side moving away more rapidly, and
so on. It is this same picture, but in three dimensions, that
we can imagine within the expanding Universe.

On this picture the velocity V of movement away from us of an
object of distance R is given by a simple law

$$V = R/T$$

where T is some time constant of the Universe.

The evaluation of this constant T depends on detailed astronomical measurements not only of the Doppler effects of distant
galaxies (which determine V) but also on measurements of their

distances away (to determine R). The distance estimates are by
far the more difficult to obtain accurately. In his book *The
Realm of the Nebulae* published in 1936, the astronomer
E. P. Hubble summarized observations up to that time and gave a
figure for T of 1800 million years. This value was accepted for
many years until the great astronomer Baade showed in 1952 that
the distance measurements had previously been seriously under-
estimated and that the true value for T was considerably greater.
The current value of T which is widely accepted as being reliable
is 18,000 million years.

The significance of this "time constant" of the Universe may be
realized if we assume that the distant galaxies are moving away
from us at uniform speed and are not being accelerated. For
then we can project back into the past to see how close galaxies
were together at various stages in history. On carrying this
to the limit we find that all the galaxies of the Universe must
have accumulated into a small region 18,000 million years
ago. The whole behaviour of the Universe would then appear as
if an explosion or "big bang" had taken place at that time in
this concentrated space and that the material of the Universe
started flying apart. Over the ensuing millions of years the
faster moving matter would cover greater distances than the more
slowly moving fragments, leading to an expanding Universe consistent with what is presently observed.

Should we say, therefore, that 18,000 million years is the age
of the Universe? It seems surprisingly short when we consider
that the lifetime of a typical sun is about ten thousand million
years, and on all plausible arguments for how galaxies have
accumulated and formed we arrive at necessary time scales greater
than this. Even within our own galaxy we can see stars that are
old and which have run through their life cycle, as well as
others which have finally reached an explosive *super nova* stage.
We would like to imagine that our own galaxy has had more time
to evolve in peace than would have been granted it by a "big
bang" 18,000 million years ago.

Some scientists do indeed believe that the Hubble constant T
should be taken to be the "age of the Universe" and that everything started with a "big bang" at some such time in the distant
past. Yet we shall see next that this is not a necessary conclusion and that the Universe is perhaps even of "infinite age"
and that the question *"How old is the Universe?"* may therefore
be meaningless.

The steady state theory of the Universe

With our knowledge that the distant galaxies are moving away
from us and that the Universe is expanding, it seems that we
cannot escape the conclusion that the Universe is changing with
time. The rate of passage of time itself may be changing in a
manner which is "tied in" with the overall changing of the
Universe. It seems at first sight that something like the "big
bang" theory of the Universe is inescapable.

In such a theory, of course, nothing can be said at all about
how all the matter of the Universe was originally created. This
thought provided the motivation for an ingenius theory regarding
the Universe proposed by Hoyle, Bondi and Gold. Briefly, the
basis of this so-called "Steady-state" theory is that, if we
must assume that matter was somehow created anyway, why not
assume that it is going on continuously? Detailed calculations
show that if one neutron or hydrogen atom "popped into existence"
so rarely that it need occur in 1 litre of space only once in
500,000 million years, this would be enough to keep the density
of the Universe constant.*

As existing galaxies draw further apart over the thousands of
millions of years, so new matter would gradually come together
and, under the action of gravitational forces, new galaxies
would continually be forming. In this way the average distance
between galaxies throughout all time would remain constant even
though galaxies themselves are being formed, go through an
evolutionary period, grow old and gradually become "dormant".

Philosophically this theory has considerable attraction as it
would mean that the Universe would, on the average, be remaining constant. The cosmological principle would apply not only
to space but also to time. An "observer" located anywhere in
the Universe and at any time would automatically see the same
sort of universe around him. He would always see distant
galaxies moving away from him, but the average distance between
galaxies would always be the same. How the matter comes into
existence is, of course, far beyond our understanding by any
present laws of science; yet it requires no more of an assumption than to say that all the matter of the Universe was suddenly

---

*This means that only about 30 atoms would have been created
within the volume of the largest of the Egyptian pyramids,
since it was built over 9000 years ago!

created at some time in the distant past, because we equally
well have no means of knowing how that could have occurred.

An advantage of this theory is that at least it can be tested
by observation.

As new telescopes and radio telescopes are being built, more
and more information can be obtained about galaxies more and
more distant in the Universe. Already some information is being
obtained about the Universe ten thousand million light years
away. But this information does not apply to what the Universe
may be like out there *now*; the very method of stating the distance means that the radiation has taken ten thousand million
years to reach us. Thus we are already looking back into the
past at a sample of the Universe as it was ten thousand million
years ago - a time that is almost as large as the Hubble time
constant itself.

If the Universe is really changing according to a "big bang"
type of hypothesis, there is no doubt that ten thousand million
years ago the matter of the Universe should have been closer
together, and in particular any galaxies which had formed should
have been closer together than we observe them today from Earth.

If with these observations it is found that the galaxies ten
thousand million light years away are very much closer together,
the "Steady-state" theory may well have to be abandoned; conversely, it would receive further strength if a sample of the
Universe ten thousand million years ago looks, on the average,
the same as we see around us today.

So far such a direct measurement of the densities of galaxies
has not provided a clear-cut decision between big bang and
steady state theories, but other less direct evidence, relating
to the general distribution of electromagnetic radiation in the
Universe, at present strongly favours the big bang theory. On
the other hand, if the "Steady-state" theory in the end should
be found to hold, there could be no question as to when the
Universe started nor could there be a question as to how old it
was. The Universe would simply be infinite in space and time,
continually expanding, and with new matter coming into existence
to maintain a constant density.

Changing time scales

The different theories of an expanding and a steady-state universe lead to another and very surprising consequence, which

requires first a discussion of another fundamental physical
concept, that of mass.

It is usually taken that the fact that the fixed stars appear
to rotate about the Earth's axis once a day, is really evidence
for the fact that the Earth rotates once a day about its axis.
But would we be led to different conclusions about anything in
the Universe, if instead we assumed that the Earth was stationary
and all the stars rotated about it?

This is a question which Newton himself wondered about and which
has occupied the attention of philosophers and scientists ever
since Newton's day. The great Viennese physicist and philosopher
Ernst Mach (1838-1916) devoted a tremendous amount of attention
to this question and he eventually concluded that there was no
experiment that could distinguish between the two situations.
This led him to suggest that the distant stars and galaxies in the
Universe must somehow be exerting an influence on the laws of
mechanics here on Earth. Just as our sky is dark because of
properties of the distant galaxies, so perhaps the entire frame-
work of mechanics as we know it, with objects having mass and
moving under the influence of forces according to Newton's laws,
is determined in some manner by an influence from the far-reach-
ing galaxies of the Universe. This in turn would mean that the
mass of any object must be influenced by the distribution of
all matter in the Universe.

Now, if the Universe is expanding, we would expect that in
another thousand million years, for example, the distant galaxies
will be further away from us than they are now - in other words,
that the density of the Universe will decrease. If the mass of
any object is determined by all the other matter in the Universe,
it is natural to expect on the above picture that the masses of
objects will gradually be changing. This may mean that the rate
of passage of time is also gradually changing. As we have seen,
we measure the passage of time by means of what we call clocks
and any repetitive device such as a planetary orbit described
under gravity or something that oscillates regularly can be
used as a clock. It is, however, impossible to compare intervals
of time which occur at completely different periods in the
Universe's history so that we cannot automatically say, for
example, that time must be "flowing uniformly".

It may be that *dynamical* clocks, such as those which employ the
period of motion of a planet in orbit, are ever so gradually
changing their rates due to changing masses.

On the other hand, we recall from Chapter 3 that we also have so-called *atomic clocks*, which are dependent on the fundamental periods associated with the movement of electrons within atoms and molecules. If masses are gradually changing, what about the electric charges of electrons and protons? The fundamental unit of an electric charge - the charge of an electron - is something entirely different from mass. Thus with the passage of time, atomic clocks may be changing their rates quite differently from dynamical clocks. If a dynamical and atomic clock are in perfect synchronism today, it could be that in a thousand million years' time the seconds or minutes which each ticks away will be quite different.

In an expanding Universe it is thus possible that different phenomena work on different time scales. It would be possible, for example, that a thousand million years ago the Universe was, as it were, much more "wound up" so that much more action occurred in what we now call one second; it may also be that, in this sense, the Universe is gradually "winding down". All this, however, is mere speculation; the definite point to be made is that masses may be gradually changing while charges may not, and that one cannot automatically think of "a uniform flow of time".

Chapter 8

TIME IN NATURE

Geological time

So far we have been concerned essentially with the nature of time, and we have seen that there is still much here that is uncertain. Nevertheless, it would be wrong to give the impression that this lack of full understanding should prevent us from using the concept of time in scientific investigations, and we now turn to the relationship of time and natural events.

In the last chapter we discussed the question of the age of the solar system, and the related question of the age of the Earth has fascinated men for a very long time. For most of this time it has been bound up with religious faith. In the sixteenth century, Bishop Ussher, on the basis of the Bible, had fixed the creation of the earth at 4004 B.C., and while there was some argument as to whether this figure was exactly correct, no one doubted that it was of the right order of magnitude. This is shown rather dramatically by the fact that when a hundred years later, Newton calculated the age of the Earth from the time it would take for it to cool from red heat to its present temperature, he rejected his result, which was of the order of 50,000 years, becaused it conflicted with the evidence from the Bible. Similar arguments led Buffon in the latter half of the eighteenth century to calculate the formation of the Earth to have occurred in 73,058 B.C. (the five figure accuracy is endearing!). By then the age of enlightenment had replaced the age of religion, and the authority of the Bible no longer carried the weight it had before. Nevertheless, it was a long struggle before it became universally, or almost universally, accepted that the evidence of nature pointed to an Earth much older than that described in the first book of the Bible. And while the original calculations were based on the existing knowledge of physics, the confirmation that the Earth was old indeed came from the rapidly accumulating geological evidence. The periods of time, which came to be believed to have elapsed since the creation of the Earth, grew longer and longer, so that they at first changed man's concept of time and eventually became so

75

huge as to pass man's imagination. What are we to make of the
statement that the Earth is now believed to be about 4500
million years old, and that life has existed on it for more than
2000 million years?

Before answering this question, we should give some indication
as to how these huge periods of time have been measured. There
are a number of naturally occurring radioactive elements, one
for instance being uranium, which change through a series of
radioactive decays into lead. These changes are such that after
a definite time, particular to the radioactive decay for each
element and known as the half-life, half the atoms of a radioactive material will have decayed into atoms of another kind.
We know from laboratory measurements the magnitude of each half-
life and we can therefore calculate the proportion of, say,
uranium and lead, which would exist at any given time after a
lump of pure uranium had started to decay. An analysis of the
proportion of lead in uranium ores will therefore tell us the
time that has elapsed since the formation of the ore. Analyses
of this type have led to the kind of time periods mentioned
above. What is of interest to us is that it is assumed that
the rate of radioactive decay has remained unchanged over these
times, but since there are no other clocks with which to compare
these radioactive ones, it is difficult to see how we can verify
this assumption. It is incidentally quite a reasonable one,
since radioactive decays which occur inside the atomic nucleus
are remarkably immune to outside influences of the kind that may
have occurred since the formation of the Earth.

We return now to the question as to how we can link these enormous periods in time with our common everyday experience. The
usual way to do this is to use time compression, i.e. to imagine
that the whole development of the Earth had taken place in, say,
a year. It is then possible to say when during that year particular geological events occurred, and it is always startling to
discover how recently most of what one normally thinks of as the
history of the Earth happened, with dinosaurs appearing in early
December, mammals just in time for Christmas and man very late
on 31st December. Recorded history becomes a matter of the last
few seconds of the year on this scale. While all this may help
us to appreciate our own insignificance in the huge scheme of
things, it is doubtful whether it does much to make us appreciate
the hugeness of the scheme itself. Another way of shrinking
time consists of counting generations, and this may be of significance when we come to discuss evolution. Thus there may have
been about ten thousand generations of man since he first came

upon the scene. The same number of generations of fruitflies,
a favourite among experimental geneticists, would have taken a
thousand years, but of viruses less than a year.

One matter that is of particular interest to us is how the year,
the month and the day have changed throughout geological time.
It is known that due to tidal friction the rate of rotation of
the Earth on its axis has slowed down and that in consequence
angular momentum has been transferred to the Moon. Surprisingly,
this does not result in the Moon speeding up in its orbit, but,
as can be shown from Kepler's laws of planetary motion, in the
Moon moving further away from the Earth and a longer month. On
the other hand, there is no apparent reason why similar changes
should have occurred to the rate of revolution of the Earth
about the Sun. From this we conclude that in earlier ages both
the day and the month were shorter, but not the year, and in
particular it has been calculated that in the Devonian period,
some 400 million years ago, the year contained about 400 days.
Recently, it has been found that the banding on certain corals
could be interpreted in terms of annual, monthly and even daily
growth. When this was done, it was found that while modern
corals had indeed about 360 day ridges in an annual growth, for
those dating from the Devonian period the corresponding number
varied from 385 to 410. Further, these ridges were in sub-
groups of average number 30.6, indicating that the Devonian year
had 13.0 lunar months, as against the present 12.4 lunar months
in the year. The geological evidence therefore strongly con-
firms the calculations based on celestial mechanics.

Time and evolution

We next turn to the processes of evolution and their relationship to time. The main problem about any investigation of the
rate at which evolution has progressed lies in the criteria
which we apply to evolutionary progress. One might measure the
times that it has taken to produce different new species, but
the concept of species is not sufficiently well defined for this
purpose. In spite of this, it is clear that evolutionary processes have worked at very different rates in different instances.
Thus it has been estimated that considerable changes took place
in the honey bee over a comparatively short period about thirty
million years ago, but almost none since. How slow these changes
can be is indicated by an investigation, conducted by J. B. S.
Haldane, who is responsible for much of our knowledge in this
field. He found that the average length of certain teeth of the
horse increased by less than a millimetre in about one and a

half million years, corresponding to half a million generations.
On the other hand, there have been exceedingly rapid evolutionary changes, such as, for instance, the evolution of the human
brain over the same time span. Changes can even be observed
in historical time, particularly as the result of the inter-
ference of man in the environment. Thus several species of
moths in Britain changed from a very light colour to almost
black over the last one hundred and eighty years, almost certainly as a result of the blackening of tree bark through industrial pollution, which made it more difficult for the dark moths
to be seen by predators than the light ones. Now there is some
evidence of the reversal of this process, following the passing
of the Clean Air Act. Other, apparently quite spontaneous,
changes have occurred among bacteria, so that a disease, such
as scarlet fever, quite suddenly decreased in virulence.

An interesting point concerns the relationship of the evolutionary process to the time available for it. It is generally
assumed that, because geological time has extended for so long,
there must have been plenty of time available for evolution to
take place. An interesting calculation that this may not be so
has been presented by H. Kalmus, who considered the case of
evolution due to the mutation of genes. In order to have at
least one of every possible mutation of a single species of
fruitfly, he calculated that so many would be required that
their total mass would considerably exceed the mass of all the
stars in the Universe. Obviously, therefore, only the minutest
fraction of all possible types of any one species can ever
exist, and it is only from these that the "best available" can
be selected.

Finally, the evolutionary process seems to have a built-in
sense of direction, in that it has never been known to lead to
exact repetition and thus appears to be irreversible. The
possible reappearance of light-coloured moths in Britain would
not contradict this, as long as these are significantly different from the earlier versions. While this is not yet known,
the fact that laboratory experiments on animals in which
parallel sister colonies have been bred from a common stock
have always led to divergent results, indicates that evolution
tends to move always towards more diversity. This should not
be equated with more disorder; on the contrary, as has been
discussed in Chapter 6, biological processes as a whole, and
evolution is no exception, lead to an increase in order. While,
as has been stated, this is not in conflict with the overall
applicability in nature of the Second Law of Thermodynamics,

the statistical arguments which lead to irreversibility in
evolution may be different from those which form the basis of
the Second Law. In fact, in a more general sense, evolutionary
processes can be convergent. For example, quite different
species have evolved quite independently the means to fly, i.e.
birds, bats, flying fish, etc. This does not contradict the
divergent nature of a particular evolutionary chain.

Biological rhythms

The natural processes which we have discussed so far belong to
that aspect of time, which manifests itself in permanent change.
Rhythmic change is of equal importance and will now be discussed.

We have already referred to the fact that the periodic motions
of the rotation of the Earth on its axis, the revolution of the
Moon about the Earth and the revolution of the Earth about the
Sun are reflected in the growth pattern of certain corals.
This is beyond any doubt a rhythm, due to external causes, in
this case one induced in the corals by periodic changes in their
environment. Such rhythms, which are called *exogenous*, are
likely to be exceedingly common, since the three periodic motions
referred to generate enormously powerful periodic changes in
the environment of biological systems - day and night, the tides,
the seasons. What we wish to study is the extent to which the
observed rhythms may by now have become wholly or partly built
into the biological systems and continue even when the external
environment is unchanging. Such internally generated rhythms
are called *endogenous*.

There has been a very large number of investigations, in which
a plant or an animal has been taken out of its natural habitat
and placed into a constant laboratory environment. Under these
circumstances it has been found that normal daily rhythms frequently persist for a considerable time. On the other hand,
they do not maintain a period of exactly a day, but often fall
short of this or exceed it by an hour or two. For that reason
they have been called circadian rhythms (*circa* - approximately,
dies - day). Similar results have also been obtained in experiments on man. Thus a group of men, who spent several months in
a completely isolated cave, moved to a 27-hour cycle. It would
therefore appear that circadian rhythms do persist in the
absence of external stimuli, but may drift from the 24-hour
rhythm, unless this is externally imposed.

It must be said, however, that it is not at all easy to esta-
blish by experiment that the circadian rhythms persist in the

absence of external stimuli, since the effects of the alternation of day and night are so all-pervading. They include not
only the obvious ones of light, temperature and humidity, but
also pressure, gravity, electric and magnetic fields, and cosmic radiation. For that reason, a crucial experiment could
only be carried out in a space-ship, and it is exactly there
where the answer to this question may one day be of great importance. For if the rhythms are entirely exogenous, this might
have serious consequences for the space traveller, whose mental
and physical well-being almost certainly depends on the main-
tenance of these rhythms. It is by now after all a common
experience that the upset of one's internal timing, that arises
from long distance jet travel in the east-west or west-east
directions, can be very unpleasant until body and mind adjust
to the new local time. To be permanently in that state could
lead rapidly to serious disorders.

However, fortunately for the future of space exploration, most
investigators agree that the circadian rhythms exist and are at
least in part endogenous, but in the long run, in the absence
of external stimuli, they may fade away. This has been found
to be so in some plants which had been bred in darkness,
although it was then also found that it was possible to reesta-
blish the rhythm by means of a single stimulus, such as for
instance one brief exposure to light, after a long period of
constant darkness. The fact that a single external stimulus
can achieve this rather than that a periodically repeated one
would be needed, is strong evidence for the existence of endogenous rhythms. The previous fading away of the rhythm has
then been explained by postulating that rhythmic behaviour is
due to a co-operative effect of a number of internal clocks.
In the absence of external stimuli, these clocks get out of
step, but a single stimulus is adequate again to achieve
synchronization.

The question arises, whether the endogenous rhythms are inherited
or acquired at a very early age. It is always exceedingly
difficult to distinguish inherited and acquired characteristics,
and no absolutely firm evidence has been presented. On the
other hand, the fact that animals reared both under constant
conditions and under conditions that varied in ways that were
different from those occurring naturally, have exhibited circadian rhythms, favours the inheritance theory. What cannot be
entirely excluded is the possibility that the rhythms were
acquired environmentally through natural variations, such as
those of cosmic ray intensity, which remained constant throughout

the experiments. It should be noted, however, that Paul Fraisse,
one of the greatest authorities in the field, states quite
unequivocally in his book (*The Psychology of Time*, p. 27) that
"These rhythms are acquired. Nothing proves this better than
a study of the development of a child. No form of circadian
rhythm is to be observed in the foetus or the new-born child,
either in its activities or in its physiological functions."
As so often in the thorny problem of heredity versus environment,
"You pays your money and you takes your choice."

One of the most significant aspects of biological rhythms, and
one that is far from fully understood, is their relation to body
temperature. It has been found that when subjects whose body
temperatures varied from 36.5°C to 39.5°C were asked to count
seconds, the frequency of counting as a function of temperature
broadly followed the general law of chemical reaction kinetics.
On the other hand, this variation was much less than would be
expected on the basis of the temperature variation of metabolic
processes, which are exceedingly temperature dependent, so that
it would appear that biological clocks are not basically metabolic in their nature. This conclusion is strengthened by the
observation that normally hibernating animals will go through
their usual hibernation cycle, during which the body temperature
varies from around 37°C to near the freezing point, even when
the external temperature is kept artificially constant. On the
other hand, it has also been found that the fiddler crab, whose
appearance changes from light to dark and back again every
24 hours, will maintain this cycle when cooled from 26°C to 6°C,
but that when the temperature is dropped to 0°C, the change is
arrested and does not recommence until the temperature is raised
again.

Even more intriguing is the relationship of the rate of the
rhythms to the intensity of ambient light, even when this is
kept constant. It is then found that some rhythms accelerate
in bright light, while for others the opposite is the case. In
general, this seems to correlate with whether an animal is nocturnal or day-active, the former having their rhythms slowed
down by constant bright light, while the latter have them speeded
up. This feature is connected with the ability of animals to
maintain a regular daytime activity in spite of the substantial
variation of the length of the day in temperate regions. Thus
it is found that a nocturnal animal synchronizes its internal
clock at dusk. In spring, as days get longer, the intensity of
the light increases and slows up the clock, so that the animal
wakes up later in the evening. The opposite happens in autumn.
Similarly, day-active animals synchronize at dawn.

One of the most obvious rhythms in man is the monthly rhythm of the female menstrual cycle, and there is also evidence for monthly male hormone rhythms, which may account for corresponding psychological rhythms which have been observed and to which the expression "lunacy" refers. There is no satisfactory explanation as to why these rhythms should be monthly, but it is possible that they are connected with the fact that our ancestors were defenceless against attack by predators during sexual activity, which was therefore safest at night during a new moon. Recently, it has been proposed by Sir Alister Hardy that man may have had a long sea shore period during his evolutionary process (see also Elaine Morgan, *The Descent of Woman*, Souvenir Press, 1972). Since the tides are closely associated with the phases of the Moon, there may be a connection here to the menstrual cycle.

Evidence has also been obtained for endogenous rhythms of an approximately annual period - circannual rhythms, and we have already mentioned the fact that hibernating animals can go through their annual cycle, even when kept under constant temperature conditions. The evidence whether man too has a circannual cycle is inconclusive.

A phenomenon which is likely to require both circadial and circannual rhythms is the migration of birds. The mechanism by which birds navigate is still not fully understood, but it is now clear that at least in part it is done with the help of both the Sun and the stars. Thus when starlings are kept in a cage, they indicate by their behaviour in which direction they wish to migrate, but they stop doing this when the sky is heavily overcast. Further, it was possible to deceive the birds by shielding them from the direct Sun and allowing them to see the Sun reflected from a mirror. Since the position of the Sun changes during the day, direction finding by the Sun inevitably necessitates knowledge of the time of day, so that birds must be able to use their circadian rhythm as a clock. Some birds have also been shown to be able to navigate by the stars, which was done by releasing them inside a planetarium. Finally, there is evidence that when migratory birds are captured and displaced a considerable distance from their normal path, they have the ability to allow for this and, when released, fly in a line directly towards their destination. This would require navigatory powers of a very high order.

Evidence for the use of the circannual rhythms in migration has come from birds that have been kept in constant conditions both

in their summering and wintering areas. These showed the same
kind of restlessness at the normal migratory times as birds
kept in natural surroundings.

While the evidence for the existence of biological clocks in
plants and animals, including man, is overwhelming, their exact
physiological location is however quite uncertain. Endogenous
rhythms have, for instance, been observed in rats that had been
deprived of most of the bodily organs, including the stomach,
adrenals, pituitary and thyroid glands, cortex, and corpus
striatum of the brain. It is difficult to understand this,
unless the clock mechanism is a property of every cell, and
studies on cells in tissue culture, which have revealed rhythmic
metabolic activities, strengthen this conclusion which, if
correct, might well account for the great firmness of our sense
of time.

Chapter 9

TIME AND MAN

The child's developing view

In the first chapter we attempted to look back on our own
experience in order to discern how our ideas of time had
developed. The purpose of the first section of this chapter is
to base the study of this development on a firm foundation of
experiment. The work described is due almost entirely to
Jean Piaget, to whom we owe so much of our knowledge of the
mental development of children.

Throughout this book, we have been unable to avoid spatial
similes, when discussing time, and the question arises, whether
this is a clue to a deeper knowledge, as to how we approach an
understanding of time. Now we can think of spatial relations
in a static way as a collection of objects simultaneously perceived in certain positions, but we cannot think of time relations in a similar way, because we cannot change the above statement with a few word alterations into one regarding time relations. The reason for this is that the perception of time
involves change in the spatial relationship of objects. In
other words, it is possible to conceive spatial relations without an extension in time, while it is not possible to conceive
temporal relations without an extension in space or, as Piaget
has put it: "Space is a still of time, while time is space in
motion." This means that we can often ignore time when describing patterns in space, but we cannot ignore space when describing patterns in time. As mentioned in Chapter 4, the link
between the two is motion, i.e. velocity, and we conclude that
an understanding of time cannot be achieved without a corresponding understanding of velocity, and conversely.

The conclusion which we have reached is closely similar to that
reached in Chapters 4 and 5, where we discussed Einstein's Theory
of Relativity. There, starting from the operational definition
of time, we were led naturally into a discussion of the concept
of velocity. The similarity is not accidental, for Piaget
records that his work started from a series of questions put to

him by Einstein. What is remarkable is that these questions, which arose from the Theory of Relativity, have in fact resulted in a profound increase in our knowledge of how children first perceive time. We are also back again with the circular argument that velocity needs an understanding of time, and time an understanding of velocity. In relativity theory we get over this by postulating the constancy of the velocity of light, in terms of which correspondences between spatial and temporal relations can then be established (see end of Chapter 5). We met the logical conclusion of this in astronomy, where we measured distances in light years, i.e. the time taken by light to travel the distances, but this point of view is indeed common in primitive people. Thus an eskimo, asked how far off a distant landmark is, may reply "many sleeps".

To find out how children perceive time, Piaget devised a series of experiments. While they are essentially simple, they do not lead as simply to clear-cut conclusions. However, the validity of the eventual conclusions is considerably increased by the number and diversity of experiments of which the following one is typical. If liquid is poured from one vessel into another, then two simple motions are involved - a drop of level and a rise of level. The child is shown the liquid being poured, the vessels being of different shape. He is given prepared forms which show the outlines of the two vessels, and asked at various times to mark on them the levels reached. The forms are then shuffled and he is asked to put them into the right time order. When this has been achieved, perhaps with help, each form is cut so that each piece of paper now contains the outline of only one flask. It is now possible to ask questions relating to simultaneity, ordering in time and duration of time. It is then found that children do not necessarily relate the ordering process - A is before B, B is before C - to the duration process - AB is shorter than AC. Now the ordering process in itself does not really relate directly to time, but merely to succession; it is the duration process which relates to time and which depends essentially on velocities. It is extremely difficult to abbreviate Piaget's accounts of his experiments without loss of clarity, and all that has been attempted here is to indicate the kind of experimental situation that eventually led him to the conclusion that children do indeed have an intuitive* feeling for velocity before they have a corresponding feeling for

---

*Piaget describes the intuitive stage as one in which a child begins to handle concepts abstractly, although thinking is still tied closely to the perception of concrete events.

duration in time. In fact, velocity is perceived first through
one thing overtaking another, in other words through ordering
and not through duration. At this point it may be objected that
a baby's first experience of time is indeed one of duration,
e.g. the time between feeds. Piaget concedes this, but says
that this is a sensory-motor experience, that is an experience
where a sensation (hunger) produces a reaction of movement
(crying), and that each experience has a time of its own. When
the child changes from the sensory-motor level to that of
thought, he reinterprets what he has previously learnt, and it
is then that the evidence shows that the linking of the separate
experiences is achieved through an intuitive realization of the
concept of velocity.

If time is first perceived through velocity, then this can only
happen in combination with a perception of space. What we are
saying is that children appear first to think of time operationally as

$$\text{Time} = \frac{\text{Distance}}{\text{Velocity}}$$

and not of velocity as

$$\text{Velocity} = \frac{\text{Distance}}{\text{Time}} \; .$$

The relationship between these two equations takes a long time
to perceive. A possible reason for this necessity of interpret-
ing time through space is that while, experientially, time always
moves in one direction, thinking about time requires us, from
the vantage point of the present, either to retrace it into the
past or to project it into the future. The experience of moving in one of two directions is of course a familiar one in
spatial relationships and it is taken over into the thought
processes of time, through the intermediary concept of velocity.
Time duration is then perceived through spatial separation,
combined with an intuitive feeling for velocity, based originally
on spatial ordering. For a quantitative measurement of time,
this is however not enough, for this requires the concept of
uniform velocity, which cannot be obtained simply from spatial
ordering. This indeed is where the logical circle of time →
velocity → time reappears, for uniform velocity can only be
defined through equal distances being covered in equal time
intervals. Fortunately, our learning processes are not logical,
and it is found that children acquire a feeling for uniform
velocity and for time measurement simultaneously. Nevertheless,

the result lacks logic, and further, the concepts of both velocity and time rest ultimately on spatial considerations. No
wonder that when we come to reflect rationally at a later age
on our concept of time, we get confused, and that it took an
Einstein to sort out the confusion. Could it be that Einstein,
who rejected the processes of formal education and in some ways,
it has been suggested, was in the best sense perhaps very child-
like, was able to draw on his childhood experiences in a way
that others, who were influenced more by the formal education
which they received, were not?

So far we have been concerned with time as perceived through
external events. In parallel with this, there is a development
of what may be called psychological time, i.e. time as perceived
through the duration of inner events. It is obvious that the
development of psychological time involves physical time, but
it leads to more, for it is after all a common experience that
the same time duration as measured by external events may appear
long or short to us according to our state of mind. While it
is clearly difficult to quantify psychological time, Piaget has
suggested that the simple numerical relationship that exists
between musical intervals is an indication of a quantitative
aspect of psychological time. Another might be the ability of
some musicians to maintain absolute pitch. These intervals had
of course been established long before their numerical relationship had been discovered. However, considerations like these
clearly take us to time as seen by the adult, a subject to which
we now turn.

The adult mind

We have reached a paradoxical point in our enquiry. If Piaget
is right, then the young child who interprets space and time in
terms of velocity, is much closer to the ideas of the Theory of
Relativity than is the adult for whom these ideas are quite
strange and are difficult to comprehend. We have suggested that
this may have arisen through the teaching which is given to
children through formal education by adults who, in turn, had
been similarly taught. It is noteworthy, however, that an
interest in the time of the physicist, both Newtonian and
Einsteinian, has increased in psychiatric patients over the past
thirty years, triggered off possibly by the way time has entered
the realm of popular science and science fiction. Perhaps what
might be termed Einstein's rediscovery of the child's point of
view is leading to conflict situations in the non-Einsteinian
adult.

However, these are speculations which it would not be sensible
to pursue. Not surprisingly, the study of the mind, which may
be defined as the total of experiences and processes active in
the brain, and its relation to time has taken its starting
point not from physics, but from biology. We saw in the last
chapter that man is subject to inbuilt rhythms, circadian,
perhaps also circannual, and others, and that these enable him
to orient himself in time, even in the absence of external
stimuli. This is evidenced by a number of facts. While we have
stated that men do not maintain an accurate time sense indefin-
itely when placed in very uniform surroundings, they nevertheless keep it for a long time and to an extent that is often
astonishing. Then again, the facility that many people have to
judge the time, when suddenly woken up, and to wake up at a
predetermined time, points in the same direction.

On the other hand, there are agents that disturb our sense of
time, for instance, as we saw in the last chapter, changes in
body temperature. We also stated that these changes are even
more profound in their effect on chemical reactions that take
place in the body. One of the most important of these that
affects the activity of the brain is that of biological oxida-
tion in the nerve cells, the rate of which decreases rapidly
during childhood and then more slowly during adult life. This
might result in an apparently faster passage of time with
advancing years.

This experience, which is often noted by adults, that time
appears to pass more quickly for them than for children, may
however be related to various social, psychological and physical
differences, and in particular may be due to the fact that adults
do not normally acquire new information and impressions at as
high a rate as they did when younger. On the other hand, when
they do, as for instance when they come to live in a new country,
then for a short period they are again like children and time
passes more slowly, i.e. what in fact happened in a few days
seems to have happened over a much longer period. It would of
course be absurd to deny the existence of psychological factors
quite generally in our experience of the apparent rate of
passage of time; all we are saying here is that physiological
factors may also be active. We all know how quickly time can
pass when we are interested and how slowly when we are bored.
On the other hand, similar variations in the apparent rate of
the passage of time can be produced through hallucinogenic
drugs, such as LSD and mescaline, which may indicate a close
connection between psychological and physiological factors.

An important aspect of the mind from our point of view is the
unconscious. The existence of a part of the mind of which in
itself we are not aware, but which interacts with our conscious
mind, was one of the great discoveries of modern psychology.
The unconscious has the important feature that it works differently from our conscious mind, as its successive interactions
with the conscious mind indicate. These interactions are manifested through apparently unrelated invasions of our conscious
processes, which often, however, have a remarkable and revealing
logic of their own, as is evidenced for instance by what are
called "Freudian slips", by which a person may, through a slip
of the tongue, reveal matters about himself that he consciously
would not have wished to reveal. The unconscious also manifests
itself in dreams, and perhaps most interestingly in sudden
flashes of insight, which are the essence of the creative
process.

There is much evidence, particularly from dreams and mystical
experiences, that the working of the unconscious is not governed
by the same time discipline as is the conscious; in fact Freud
stated that in the unconscious there is no before and after.
It is interesting that some of the evidence for this statement
comes from the experience of mathematicians. The conscious
thought processes in mathematics are unidirectional not only in
time but also in logic, and are therefore particularly suitable
for study from our point of view. Now, again and again, mathematicians have recorded their experience of the sudden flash of
insight. This commonly comes quite unexpectedly, after a long
process of conscious thought on a problem had not led to its
solution. The flash of insight then suddenly presents the whole
solution in a way quite unconnected with the previous conscious
thought processes, and all at once, rather than over a time
interval in logical order. The experience is of course not confined to mathematicians, but forms a crucial part in the act of
creation in any scientific activity and in many non-scientific
activities. One of the authors, for instance, has experienced
it in connection with the translation of poetry. Its most
important characteristic is the simultaneity with which the com-
plex solution arrives in the conscious mind, which is an indication of the timeless nature of the unconscious. Probably the
most famous instance of a mathematical kind has been recorded
by Poincare (see P. E. Vernon (ed.) *Creativity,* Penguin Books,
1970, p. 77).

The timelessness of the unconscious has also been linked to the
question as to whether it is possible to foretell the future.

There is, for instance, an impressive list of well-documented
dreams that relate to events that eventually turned out to have
been in the future. They range from the trivial to the trau-
matic and in many cases fraud seems excluded and any explanation
in terms of random coincidences would appear to be more far-
fetched than one in terms of the dreamer's ability to be able
to predict the future. A good collection is given in Priestley's
book *Man and Time*, and one of the curious features of these
cases is that in some instances they confined themselves to a
prediction of the future, while in others they made it possible
for the person involved to alter the future. A good example of
the latter kind is a dream in which a person kills a child who
had suddenly run in front of his car, and who later experiences
the same situation in every detail in real life, except that,
having been forewarned by the dream, he is able to stop just in
time.

Phenomena of this kind are related to the whole problem of extra-
sensory perception on which there is no agreement. Here we
would like to quote Priestley at length (*Man and Time*, p. 194):

> "There may exist a few superhumanly disinterested
> intellects, but I believe all the rest of us
> come down on one side of this fence or the other.
> In our secret depths, wherever we do our un-
> spoken wishing, either we want life to be tidy,
> clear, fully understood, contained within
> definite limits, or we long for it to seem
> larger, wilder, stranger. Faced with some odd
> incident, either we wish to cut it down or to
> build it up.
>
> On this level, below that of philosophies and
> rational opinions, either we reject or ignore
> the unknown, the apparently inexplicable, the
> marvellous and miraculous, or we welcome every
> sign of them. At one extreme is a narrow in-
> tolerant bigotry, snarling at anything outside
> the accepted world picture, and at the other
> is an idiotic credulity, the prey of any glib
> charlatan. At one end the world becomes a
> prison, at the other a madhouse. Now we may
> easily avoid these extremes; but I believe the
> secret bias to be always there, however much
> we may pretend to be disinterested and objec-
> tive; and its influence is always felt when

> something like precognition is being examined
> and discussed. We are all at heart on one
> side or the other, *wanting* it to be true or
> hoping that it is false."

We agree with Priestley that we are all biased in this matter.
His bias is indicated by the stress on the word "wanting", and
although ours is in the opposite direction, we do not accept
that "we hope that it is false". On the contrary, we hope that
means may be devised which will give us a better understanding
of these phenomena if they exist, or a clearer demonstration
that they do not. Where we have a bias is that we hope these
means will be integrated into the natural sciences, however
much these may have to be extended for this purpose.

On philosophy and religion

The reader may wonder at the few pages that follow the rather
pretentious title of this section, but the stress here should
be on the little word "on". What we are going to attempt is
certainly not a comprehensive account of the way that these
great disciplines of the human mind have approached the problem
of time. A good account, at least from the European angle, may
be found in *The Discovery of Time* by Toulmin and Goodfield.

To see how the philosopher's approach differs from that of the
scientist, we turn to Newton who stated that "In philosophical
disquisitions, we ought to abstract from our senses, and consider things themselves, distinct from what are only sensible
measures of them". It is this extension of our enquiry beyond
what is measurable and directly available to our senses that we
referred to at the end of Chapter 2, when we stated that there
are aspects of our enquiry outside the realm of science. These
aspects must not, however, ignore the positive findings of
science, nor the restrictions on any form of enquiry, based on
these. It is therefore disquieting that there is no reference
to Einstein or his Theory of Relativity either in the book by
Toulmin and Goodfield, or in that part of Fraser's *Voices of
Time* that deals with philosophy and religion, although this
fact is at least recognized at the end of the first article in
Fraser's book.

In all fairness, it must be said, however, that not all philosophers would subscribe to Newton's dictum. Thus Samuel Alexander,
a modern philosopher, has stated that "philosophy proceeds by
description; it only uses argument in order to help you see the

facts, just as a botanist uses the microscope". But if that is
so, in what sense is philosophy different from science, which
certainly uses both argument and microscopy?

Throughout history, from Heraclitus to Heidegger, philosophers
seem to have been concerned with two different dichotomies
relating to time.* The first is that of change and permanence.
To Heraclitus, it was change that was real, as is evidenced by
the famous saying attributed to him that "all things flow",
while constancy was merely the static instant of dynamic change,
and thus apparent. In contrast, Parmenides and Zeno argued that
change implied the future existence of something that did not yet
exist. But if it did not exist, then when it came to exist, it
would come from nothing, and something could not come from
nothing. Hence change was not real. The story is told that
when Zeno tried to convince Diogenes that change and hence motion
did not exist, the latter silently got up and walked away, having
thus made his point.

In modern times, the dichotomy reappears in, for instance,
Bergson's view that time takes two forms. The first is that of
duration, "which the succession of our conscious states assumes
when our ego lets itself live, when it refrains from separating
its present state from its former states". This form embodies
change. In the second, "we set our states of consciousness
side-by-side in such a way as to perceive them simultaneously,
no longer in one another, but alongside". This is not an easy
passage to understand, but to the extent that it suggests that
states of consciousness do not go over into one another, it
implies permanence.

It is interesting that the same dichotomy appears in classical
Indian thought, but that there time is viewed essentially in
terms of static permanence. While to Heraclitus it was the
flow of the river that was reality and denoted change, to
Indian thought it is the river itself which with its unchanging
pattern is static and permanent. This attitude is reflected in
Sanscrit language which, for instance, does not distinguish
between "to become" and "to exist". Altogether, the great
diversities that exist in the structures of different languages
have had a profound effect on philosophical attitudes and
conclusions.

To the extent that time displays the properties of both change
and permanence, it would appear that any attempt to come down

---

*Dichotomy: Division into two (*Concise Oxford Dictionary*).

in favour of one rather than the other is likely to be unsuccessful. But the statement that time does display two such contradictory features is of course itself paradoxical or at least was so until the famous physicist Niels Bohr drew attention to the fact that this property, of fundamental concepts and entities displaying self-contradictory features, was indeed common. It first came to the notice of physicists in atomic physics, when the electron turned out to have contradictory properties of both particles and waves, an observation which led Bohr to his principle of complementarity, according to which properties that in our common sense experience are contradictory, at a deeper level of understanding are complementary ways of describing the same thing. He has suggested that the principle might usefully be applied to such long standing dichotomies as the problem of predestination and free will, and it can equally usefully be applied to the problem of change and permanence in time.

The second dichotomy concerns the question as to whether time is relative or absolute; in other words, whether it can exist only in association with events, or has an independent existence of its own. Newton concluded that they both existed. His full statement, of which we have already quoted part, is "Absolute, true and mathematical time, of itself, and from its own nature, flows equally without relation to anything external, and by another name is called duration; relative, apparent and common time is some sensible and external measure of duration by means of motion, which is commonly used instead of true time." Thus in Newton's view there are two quite distinct times, which of course is very different from saying that we are witnessing here complementary properties of one entity.

The problem of absolute and relative time is already apparent in Aristotle's work and is still present in that of Samuel Alexander. It basically arises from the realization that any study of time must involve the measurement of time and the idea that measurement involves the comparison with an absolute standard. Here again, modern physics must come to our aid, and we know from the considerations which led Einstein to propose his Theory of Relativity, that this last conclusion is indeed false, and that there are no time standards which are independent of an observer.

It may be asked at this point how it comes about that we all appear to have much the same time standard when each standard is individual to its observer. The physicist's answer to this

is that these standards may indeed be the same, provided that
they can be operationally compared. This is possible, if two
observers are in close proximity to each other and remain so for
some time, which is not a bad way to describe our existence on
this Earth. The standards cannot be compared, according to
Einstein, when this condition is not satisfied. Of this, we
have so far little direct experience, although a caesium clock
was recently taken up in a jet aircraft (see Chapter 5). What
is important here is that the idea of absolute time is an extra-
polation of our Earth-bound experience, for which there is no
direct evidence, and we now believe that it is an injustified one.

We hope that this digression illustrates the main point that we
have been trying to make, namely that there is much in modern
physics that illustrates the age-old problems of philosophy and
that if philosophers wish to go beyond physics, they must at
least take adequate notice of where physics has got to. (See
the above remark about the absence of references to Einstein.)
It does not mean that there is nothing beyond physics. Michael
Whiteman writes:*

> "We have to thank the progress of the natural
> sciences for a successively clearer distinction
> of the genuine phenomena and laws of physics
> from myth, superstition and baseless conjecture.
> Similarly we have to thank the accompanying
> growth of the scientific spirit for a succes-
> sively clearer distinction of realities of the
> personal and inner life from confusions and
> counterfeits."

But just in case scientists should get too certain of themselves,
let us remind ourselves that the study of thermodynamics, which
at one time was thought to give a logically consistent view of
time, has failed to do so. Whitrow, in his excellent study on
The Natural Philosophy of Time, concludes that time may indeed
be an ultimate and irreducible concept. But, he says, "this
does not commit us to the unnecessary hypothesis that it is
absolute, for moments do not exist in their own right but are
merely classes of co-existent events. Nor is time a mysterious
illusion of the intellect. It is an essential feature of the
universe."

─────────────────

Philosophy of Space and Time, Allen & Unwin, London, 1967,
   p. 24.

Religious attitudes to time have always been concerned with the
problems of death, life after death and the meaning of life.
Thus the theologian Paul Tillich starts a sermon on time with
the words:* "Let us meditate on the mystery of time". Not "let
us explain" or "let us unravel", but "let us meditate on", and
it is here that we are in a realm to which science cannot con-
tribute, but which is very real. He goes on:

> "The life of each of us is permeated in every
> moment, in every experience, and in every
> expression, by the mystery of time. Time is
> our destiny. Time is our hope. Time is our
> despair. And time is the mirror in which we
> see eternity."

Regrettably, the most immediate and most obvious fact about time
in its relation to our life on Earth is that it is destructive,
for whatever we may have achieved in this life, in the end death
awaits us. Father Time not only carries an hourglass but also
a scythe.** Different religions have tackled the problem in
different ways. In some the present life is a preparation for
the next, whether this is through reincarnation on this Earth
or in a heaven and hell beyond this Earth. Others have postulated a shadowy non-existence, as in the Greek Hades. A study
of the religions of the world will teach us less about time than
about why men want to study time. "Time is our hope." And few
can have expressed the struggle against time and the essential
impotence of man more poignantly than Dylan Thomas.***

> "Do not go gentle into that good night,
> Old Age should burn and rave at close of day:
> Rage, rage against the dying of the light."

---

*Paul Tillich, *The Shaking of the Foundations*, Pelican, 1962,
  p. 42.
**It has been suggested that Father Time acquired his scythe
  through a misidentification of *Chronos* (Greek for time) with
  Kronos, the most ancient of the Greek gods and father of Zeus.
  Kronos carried a scythe, which signified not only that he was
  the god of agriculture, but also that he himself came to power
  by castrating his own father, Uranus. Of such stuff are our
  images made.
***Dylan Thomas, *Collected Poems 1934-52*, Dent & Son, London, 1952.

Time in literature

The quotation from Dylan Thomas makes a suitable transition to
our next topic, the treatment of time in literature. To show
how close the connection is with the previous topic, we should
like to start this section with a famous passage from the Bible
(Ecclesiastes, ch. 3, 1-8):

> "To everything there is a season and
> a time to every purpose under the heaven:
>   A time to be born, and a time to die;
> a time to plant, and a time to pluck up
> that which is planted;
>   A time to kill, and a time to heal,
> a time to break down, and a time to build up;
>   A time to weep, and a time to laugh;
> a time to mourn, and a time to dance;
>   A time to cast away stones, and a time
> to gather stones together; a time to embrace,
> and a time to refrain from embracing;
>   A time to get, and a time to lose;
> a time to keep, and a time to cast away;
>   A time to rend, and a time to sew;
> a time to keep silence, and a time to speak;
>   A time to love, and a time to hate;
> a time of war, and a time of peace."

Here is wisdom, as well as poetry, and to some extent it is an
answer to Thomas, in the path which the Preacher advocates
between rage and resignation.

Literature has concerned itself with time in essentially two
aspects. The first of these is the relationship between man
and time and here it supplements philosophy and religion. The
study of literature adds another dimension to this, for the art
and imagination of the great writers enables us to perceive
things which escape mere thought. Shakespeare was preoccupied
with the problem of time, and a reading of his plays and sonnets
will teach us much that is not dreamed of in our philosophy.
Modern writers who have very consciously explored the problem
of time include, for instance, Marcel Proust, James Joyce,
Virginia Woolf, T. S. Eliot and Samuel Beckett. All these have
unquestionably been influenced by the discoveries of Freud, and
they make much use of the "stream of consciousness" with its
unconscious association of ideas. This leads to a slowing up
of time that is not unlike the timelessness postulated by Freud

for the unconscious. In Joyce's Ulysses, the action and inaction
of a day is described over 700 pages and it is often difficult
to remember what time of day it is at any particular moment.
The sense of timelessness engendered is in total contrast with
the almost total time dependence of our modern life, but akin
to the Indian concept of time as something static.

It is indeed true that the art of the writer can enable the
reader to perceive things which escape thought, but such perception depends on the relationship between reader and writer,
which can be very individual and very personal. For that reason,
different readers are going to react to the passages that follow
in different ways; some may not react at all, while others may
receive an insight previously denied to them. The passages are
from the works of T. S. Eliot, probably the most conscious and
most persistent explorer of time in modern literature. Because
his poetry is so allusive and evocative and so interwoven, it
is particularly difficult to convey it through brief quotations,
and hence the purpose of the ones that follow is essentially to
send the reader to the books from which they are taken.* As
early as 1917 he writes:

> "In a minute there is time
> For decisions and revisions which a minute will reverse.
> For I have known them all already, known them all—
> Have known the evenings, mornings, afternoons,
> I have measured out my life with coffee spoons."

> (The Love Song of J. Alfred Prufrock)

In contrast with this realization of the elasticity of subjective
time is the insistent reminder of the Landlord,

> "Hurry up please it's time"

> (The Waste Land)

that, however elastic time may be, it runs out on us. And it
does not come back:

---

*T. S. Eliot, *Collected Poems, 1909-1962*, Faber & Faber, London,
1963.
T. S. Eliot, *Murder in the Cathedral*, Faber & Faber, London,
1935.

"Because I do not hope to turn again
Because I do not hope
Because I do not hope to turn"

(Ash Wednesday)

There is a curiously outdated, but probably not intended, reference to Newtonian concepts or - less likely - an unusually per-
ceptive one to Einsteinian ones,

"Because I know that time is always time
And Place is always and only place
And what is actual is actual only for one time
And only for one place."

(Ash Wednesday)

and a hint of the complementary nature of time is given in the
words of Thomas à Becket:

"It is not in time that my death shall be known;
It is out of time that my decision is taken."

(Murder in the Cathedral)

The most complete exploration, however, occurs in The Four
Quartets, undoubtedly his most profound poem which not only in
its title recalls the profundity of the late quartets of
Beethoven.* Eliot starts with a denial of the notion of time
as a straight line:

"Time present and time past
Are both perhaps present in time future,
And time future contained in time past.
If all time is eternally present
All time is redeemable."

(Burnt Norton)

---

*The third movement in Beethoven's Opus 132, entitled "Heiliger
Dankgesang eines Genesenden an die Gottheit, in der lydischen
Tonart" (Hymn of thanksgiving to the Godhead by one convales-
cing, in the Lydian mode), has been used by Aldous Huxley in
his novel *Point Counterpoint* in an attempt to persuade an
atheist of the existence of God. This movement, with its vary-
ing rhythms, is also a revelation of time, of a kind which can
only come through music.

This theme appears again in

>"In the beginning is my end."

>(East Coker)

and in

>"What we call the beginning is often the end
>And to make an end is to make a beginning.
>The end is where we start from."

>(Little Gidding)

He again refers to the complementary features of time:

>"At the still point of the turning world."

>(Burnt Norton)

>"Only by the form, the pattern,
>Can words or music reach
>The stillness, as a Chinese jar still
>Moves perpetually in its stillness."

>(Burnt Norton)

>"Time the destroyer is time the preserver."

>(The Dry Salvages)

and finally,

>"But to apprehend
>The point of intersection of the timeless
>With time is an occupation for a saint."

>(The Dry Salvages)

He echoes the Preacher in Ecclesiastes:

>"There is a time for building
>And a time for living and for generation
>And a time for the wind to break the loosened pane."

>(East Coker)

> "The time of the seasons and the constellations
> The time of milking and the time of harvest
> The time of the coupling of man and woman
> And that of beasts. Feet rising and falling.
> Eating and drinking. Dung and death."

(East Coker)

Here we are back with the brevity of life.

> "Only through time time is conquered."

(Burnt Norton)

> "Ridiculous the waste sad time
> Stretching before and after."

(Burnt Norton)

> "We, content at the last
> If our temporal reversion nourish
> (Not too far from the yew-tree)
> The life of significant soil."

(The Dry Salvages)

A reference to the first lines of Dante's Divine Comedy,

> "In the middle, not only in the middle of the way
> But all the way, in a dark wood."

(East Coker)

reminds us of the eternal search after the truth in life and time, but in the end he leaves us in hope, by quoting the medieval mystic, Dame Juliana of Norwich:

> "Sin is Behovely, but
> All shall be well, and
> All manner of thing shall be well."

If this all too brief account of Eliot's exploration of time has indicated that there are ways of exploring time that owe nothing to the methods of science, it will have served its purpose. To achieve more, it is necessary to read the poems.

The second aspect of time, treated in literature, concerns the
breaking down of the limitations that the nature of time imposes
on us. Imaginary, and sometimes imaginative, accounts of travel
in time abound, but one of the best remains H. G. Wells's *The
Time Machine*. These books serve a useful purpose in that they
explore the contradictions that arise when the normal laws that
appear to govern time are violated, so that it for instance
becomes possible for one person to travel backwards in time or
forwards at an accelerated rate, compared to the rest of mankind.
The contradictions arise when the traveller ceases to be an outside observer of the scene and attempts to rejoin the rest of
us. Thus if I travel back in time, I may look at myself taking
my degree examination, but I should be in trouble, and not only
with the examiners, if I then helped my earlier self answer the
questions, since I would thereby change my own past. This problem is of course similar to the one referred to earlier, whether
precognition that appears to change the future is possible, and
imaginative exercises like those of the science fiction writers
may one day perhaps help us to explore those realms of time
which at present appear to defy physical investigation.

The timekept city

> "I journeyed to London, to the timekept City
>
> ....
>
> I journeyed to the suburbs, and there I was told:
> We toil for six days, on the seventh we must motor
> To Hindhead or Maidenhead.
> If the weather is foul we stay at home and read the
>   papers."

This quotation from T. S. Eliot's "Choruses from the Rock" well
expresses the dilemma of modern man, who in his attempt to
master time has become its slave. In contrast, a Governor of
Ceylon wrote in 1801:* "There is not an inhabitant in this
island that would not sit down and starve out the year under
the shade of two or three coconut trees rather than increase
his income and his comforts by his manual labour."

Those who wish to return to a life free from the stresses of our
modern age should note the word "starve" in the Governor's des-
patch. Our dependence on time is intimately related to techno-

---

*Quote by John Cohen in J. T. Fraser, *The Voices of Time*, p. 273.

logical progress, and the answer to our problem does not lie in
retreating from technology, but in making technology our servant.
There is increasing evidence that technological developments
have a life of their own, so that for instance it is exceedingly
difficult to find in the United Kingdom any substantial number
of people either now or in the past who have been in favour of
building a supersonic transport plane. Yet Concorde exists and
has cost the taxpayer about £1000 million. It would appear that
here technology has made us its servant.

How to change this, so that we can live with technology while
living by technology, is not at all clear, but a clue may be
found in the way our society treats time. One of the main aims
of technology is to make commodities more easily available,
which means that we should have more of them. Now the most
obvious effect of technology on time is that this is indeed true
for many people, who no longer toil for six days, as in Eliot's
poem, but for only five, yet simultaneously time has become an
increasingly scarce commodity for others. The business execu-
tive, who, in E. M. Forster's evocative phrase,* lives in "a
world of telegrams and anger", used to cross the Atlantic by
ship in five days. Now he does it in hardly more than five
hours. Obviously, he should now take the holiday on land, that
he used to take on board ship, instead of which he plunges into
work with his metabolism upset because of the time difference
between Europe and America. Not surprisingly, he is liable to
make bad decisions under such circumstances, and hence many
firms now make a 24-hour acclimatization period compulsory, not
because it is humane, but because it is profitable.

There is a clear hint here that the tyranny of time is connected
with what may be called the gospel of work, which has dominated
Western society for so long. Although work was the curse put
upon Adam for his transgression in the Garden of Eden, "Satan
finds work for idle hands to do" used to be a common reproof.
Should there not at least also be a good angel who finds leisure
for idle hands? Let us again quote the Preacher (Ecclesiastes,
ch. 3, 9-13):

> "What profit hath he that worketh
> in that wherein he laboureth?

---

*E. M. Forster, *Howards End*, Penguin, 1941.

> I have seen the travail, which
> God hath given to the sons of men to
> be exercised in it.
>
> He hath made everything beautiful
> in his time: also he hath set the
> world in their heart, so that no man
> can find out the work that God maketh
> from the beginning to the end.
>
> I know that there is no good in
> them, but for a man to rejoice, and to
> do good in his life.
>
> And also that every man should
> eat and drink, and enjoy the good of
> all his labour, it is the gift of God."

In contrast, in Germany a man recently was successfully prose-
cuted by his employer for returning from his holiday so tired
that he was no fitter for work than when he went on holiday.
As the holiday was paid for by the employer, the court ruled
that its purpose was to refresh the employee for further work.
The writing is on the wall.

Strangely, the problem of how to occupy leisure time, so tell-
ingly described in the Eliot quotation above, is not unconnected
with the problems that arise from the tyranny of the clock. For
both arise out of the belief that there is an inherent virtue
in work, so that in our educational system it would be considered
sinful if we were prepared for leisure, which would be called
idleness. One of the main reasons why the middle-aged of today
resent the attitudes of many of the young is that the latter
have begun to reject the gospel of work and to turn for inspira-
tion to those people in the East who had never accepted this
gospel. The solution does not lie in a rejection of the one
philosophy or the other, but in the almost unbelievably difficult reconciliation of the two. In this an understanding of
man's relationship to time is crucial, an understanding which
can only be achieved in those realms of human endeavour that lie
outside the province of science.

We have come full circle - from Duchamp's picture to Eliot's
poetry, via science, philosophy and religion. All are necessary
if we are to obtain a fuller understanding of time, but none
have enabled us to find an answer to the question "What is time?"
This eluded St. Augustine nearly 2000 years ago (see title page)
and it still eludes us today.

POSTSCRIPT

The following letter appeared in the London *Times* on 11 July 1972:

A COUPLE OF TIMES

From Lady (Shane) Leslie

Sir,

A story was told by my late husband Sir Shane Leslie, about the famous Provost of Trinity College, Dublin, Professor Mahaffy. Mahaffy missed a train at a country station in Ireland. He had noticed the time on a clock outside the station which differed from another clock placed inside. Angrily he tackled an old porter on the inefficiency which had caused him to lose his train.

The old man scratched his head - then replied: "If they told the same time there'd be no need to have two clocks" - which, as Mahaffy said, was not only extremely funny but also unanswerable.

Yours sincerely,

IRIS C. LESLIE,

Old Parsonage Farm,
Hanley Castle,
Worcestershire.

8 July 1972

REFERENCES AND FURTHER READING

The list contains more than any one reader is likely to wish to read, but we have given brief notes on all the books listed, so as to make selection easier. The articles from the *Scientific American* are all worth reading, although some are not easy, and we have also listed a very few articles from other journals. (References to *Scientific American* list both the number in the reprint collection, where available, and the date and page of the original publication.) The literature on the subject is enormous, and even at the level at which this list is pitched we are likely to have missed much.

General

Butler, S. T., and Messel, H., (ed.), *Time,*
Shakespeare Head Press, Sydney, 1965, and Pergamon Press, London, 1966.

> (Goes further than the present book. Mainly on chapters 4-7.)

Einstein, A., and Infeld, L., *The Evolution of Physics,*
Simon & Schuster, New York, and Cambridge University Press, 1938.

> (Still one of the best introductions to the concepts of modern physics.)

Fraser, J. J., (ed.), *The Voices of Time,*
Penguin Press, London, 1968.

> (We have freely drawn on this excellent and very comprehensive book.)

Gold, T., (ed.), *The Nature of Time,*
Cornell University Press, 1967.

> (Mostly very advanced, but some of the discussion is very clear.)

Goudsmit, S. A., and Claiborne, R., *Time*,
   Time-Life Books, New York, 1966.

   (Very elementary. Excellent illustrations.)

Schlegel, R., *Time and the Physical World*,
   Dover, New York, 1968.

   (Definitely advanced. Mainly on chapters 4-7.)

Whitrow, G. J., *The Natural Philosophy of Time*,
   Nelson, London, 1961.

   (Not easy, but very clear. Strongly recommended for
   advanced reading.)

Whitrow, G. J., *What is Time?*,
   Thames & Hudson, London, 1972.

   (Very readable. Covers similar material to the present
   book.)

Chapter 2

Bondi, H., *Assumption and Myth in Physical Theory*,
   Cambridge University Press, 1967.

   (This and the other of Bondi's books, quoted below, show
   him to be a leading expositor in the field, easy and
   delightful to read.)

Feynman, R., *The Character of Physical Law*,
   British Broadcasting Corporation, London, 1965.

   (A brilliant and wide ranging exposition.)

Chapter 3

Lyons, H., Atomic Clocks,
   Scientific American **225**, Feb. 1957, p. 71.

Ullyett, K., *Clocks and Watches*,
   Hamlyn, London, 1971.

   (A clear and extensively illustrated account of time
   pieces throughout history.)

Chapter 5

Bondi, H., *Relativity and Common Sense*,
  Heinemann, London, 1968.

  (For comment, see Bondi's other book above.)

Bronowski, J., The Clock Paradox,
  Scientific American 291, Feb. 1963, p. 134.

Marder, *Time and the Space Traveller*,
  Allen & Unwin, London, 1971.

  (An excellent account of the twin paradox.)

Sciama, D. W., *The Unity of the Universe*,
  Faber & Faber, London, 1959.

  (Goes into considerably more detail than we do. Easy to
  read.)

Chapter 6

Coe, L., The Nature of Time,
  Amer. Journal of Physics 37, 810 (1969).

Davies, P. C. W., The Arrow of Time,
  Physics Bulletin 22, 211 (1971).

Ehrenberg, W., Maxwell's Demon,
  Scientific American 317, Nov. 1967, p. 103.

Gardner, M., Can Time go Backward?
  Scientific American 309, Jan. 1967, p. 98.

Hafele, J. C., and Keating, R. E., Around-the-World Atomic Clocks,
  Science 177, 166 (1972).

Layzer, D., The Arrow of Time,
  Scientific American, Dec. 1975, p. 56.

Morrison, P., The Overthrow of Parity,
  Scientific American 231, Aug. 1957, p. 45.

Overseth, O. E., Experiments in Time Reversal,
  Scientific American, Oct. 1969, p. 89.

Sachs, R. G., Can the Direction of the Flow of Time be
   Determined?,
   Science 140, 1284 (1963).

Trieman, S. B., The Weak Interactions,
   Scientific American 247, Mar. 1959, p. 72.

Wick, G., The Clock Paradox Resolved,
   New Scientist, p. 261, 3.2.1972.

Wigner, E. P., Violation of Symmetry in Physics,
   Scientific American 301, Dec. 1965, p. 28.

Chapter 7

Brown, H., The Age of the Solar System,
   Scientific American 102, Apr. 1957, p. 80.

Dyson, F. J., Energy in the Universe,
   Scientific American, Sept. 1971, p. 51.

Gamow, G., The Evolutionary Universe,
   Scientific American 211, Sept. 1956, p. 136.

Golt, J. R., *et al.*, Will the Universe expand forever?
   Scientific American, Mar. 1976, p. 62.

Harrison, E. R., Why the Sky is Dark at Night,
   Physics Today, Feb. 1974, p. 30.

Hoyle, F., The Steady State Universe,
   Scientific American 218, Sept. 1956, p. 157.

Kirkaldy, J. F., *Geological Time*,
   Oliver & Boyd, Edinburgh, 1971.

   (A comprehensive and easily read account of the development of our ideas of geochronology.)

Reynolds, J. H., The Age of the Elements in the Solar System,
   Scientific American 253, Nov. 1960, p. 171.

Chapter 8

Barghoorn, E. S., The Oldest Fossils,
   Scientific American, May 1971, p. 30.

Keaton, W. T., The Mystery of Pigeon Homing,
  Scientific American, Dec. 1974, p. 96.

Luce, G. G., *Body Time,*
  Temple Smith, London, 1972.

  (Rhythms in man and how modern life affects them.)

Palmer, J. D., Biological Clocks of the Tidal Zone,
  Scientific American, Feb. 1975, p. 70.

Pengelley, E. T., and Asmundson, S. J., Annual Biological Clocks,
  Scientific American, Apr. 1971, p. 72.

Rosenberg, G. D., and Runcorn, S. K., (ed.), *Growth Rhythm and
  the History of the Earth's Rotation,*
  Wiley, London, 1975, especially pp. 285-292.

Runcorn, S. K., Corals as Paleontological Clocks,
  Scientific American 871, Oct. 1966, p. 26.

Saunders, D. S., The Biological Clocks of Insects,
  Scientific American, Feb. 1976, p. 114.

Ward, R. R., *The Living Clocks,*
  Collins, London, 1972.

  (A good account of biological clocks and of the scientists
  who discovered them.)

Chapter 9

Bryant, S. W., What Jet Travel does to your Metabolic Clock,
  Fortune 68, 160 and 183 (1963).

Cohen, J., The Scientific Revolution and Leisure,
  Nature 198, 1028 (1963).

Cohen, J., Psychological Time,
  Scientific American, Nov. 1964, p. 117.

Fraisse, P., *The Psychology of Time,*
  Harper & Row, New York, 1963.

  (A very comprehensive account, but not easy to read.)

Moore, W. E., *Man, Time and Society*,
  Wiley, New York, 1963.

(A sociologist's view of how we are affected by time.)

Ornstein, R. E., *On the Experience of Time*,
  Penguin Books, 1969.

(An account of some interesting psychological experiments.)

Piaget, J., *The Child's Conception of Time*,
  Routledge & Kegan Paul, London, 1969.

(Not an easy book, but the conclusions provide an excellent summary.)

Priestley, J. B., *Man and Time*,
  Aldus Books, London, 1964.

(A very personal account by a creative writer.)

Tillich, P., *The Shaking of the Foundations*,
  Pelican, 1962, especially pp. 42-45 and p. 154.

Toulmin, S., and Goodfield, J., *The Discovery of Time*,
  Penguin Books, 1969.

(The development of the concept of time in history.)

Williamson, G., *A Reader's Guide to T. S. Eliot*,
  Thames & Hudson, London, 1967.

(The book ably explores the many-layered nature of Eliot's poetry in comparatively simple language.)

Zwart, P. J., *About Time*,
  North-Holland Publishing Co., Amsterdam, 1976.

(A philosophical enquiry into the origin and nature of time. Very clear and full of insight.)

INDEX

When an entry occurs on two consecutive pages, only the first page is given.

113